Biochemistry and Molecular Biology of Antimicrobial Drug Action

Fifth edition

Formerly

Biochemistry of Antimicrobial Action

T. J. Franklin and G. A. Snow

Zeneca Pharmaceuticals
Alderley Park, Macclesfield
Cheshire

KLUWER ACADEMIC PUBLISHERS
DORDRECHT / BOSTON / LONDON

Library of Congress Cataloging in Publication Card Number: 98-70539

ISBN 0 412 82200 8

Published by Kluwer Academic Publishers,
P.O. Box 17, 3300 AA Dordrecht, The Netherlands.

Sold and distributed in North, Central and South America
by Kluwer Academic Publishers,
101 Philip Drive, Norwell, MA 02061, U.S.A..

In all other countries, sold and distributed
by Kluwer Academic Publishers Group,
P.O. Box 322, 3300 AH Dordrecht, The Netherlands.

Printed in Great Britain

This edition is dedicated to the memory of Dr G. Alan Snow, 1915–1995, my co-author, colleague and mentor.

*The cover illustration shows a computer-generated graphic of Penicillin V *phenoxymethyl penicillin) kindly provided by Mr A.M. Slater*

Contents

Preface

The rapid advances made in the study of the synthesis, structure and function of biological macromolecules in the last fifteen years have enabled scientists concerned with antimicrobial agents to achieve a considerable measure of understanding of how these substances inhibit cell growth and division. The use of antimicrobial agents as highly specific inhibitors has in turn substantially assisted the investigation of complex biochemical processes. The literature in this field is so extensive, however, that we considered an attempt should be made to draw together in an introductory book the more significant studies of recent years. This book, which is in fact based on lecture courses given by us to undergraduates at Liverpool and Manchester Universities, is therefore intended as an introduction to the biochemistry of antimicrobial action for advanced students in many disciplines. We hope that it may also be useful to established scientists who are new to this area of research.

The book is concerned with a discussion of medically important antimicrobial compounds and also a number of agents that, although having no medical uses, have proved invaluable as research tools in biochemistry. Our aim has been to present the available information in a simple and readable way, emphasizing the established facts rather than more controversial material. Whenever possible, however, we have indicated the gaps in the present knowledge of the subject where further information is required. We have avoided the use of literature references in the text; instead we have included short lists of key articles and books for further reading at the end of each chapter.

We have drawn on the work of many scientists and we are especially pleased to express our thanks to those who have given us permission to reproduce their original diagrams and photographs. We are also grateful to the Pharmaceuticals Division of Imperial Chemicals Industries Ltd, for providing the necessary facilities for the preparation of this book.

Abbreviations used without definition for common biochemical substances are those recommended by the Biochemical Journal (1970).

June 1970

T. J. FRANKLIN
G. A. SNOW

Preface to the fifth edition

Since the previous edition of this book there have been major developments in medicine and its underpinning basic sciences. The problems posed by infectious diseases afflicting humans and their domestic animals have attracted increasing publicity and concern throughout this period. The menace of AIDS continues unabated and has reached epidemic proportions in parts of the developing world. The spread of multidrug-resistant bacteria is seemingly unstoppable, sporadic outbreaks of meningitis, bacterially mediated food poisoning and lethal viral infections out of Africa regularly alarm the public. Fortunately, against this somewhat gloomy picture can be set some notable advances. The impact of rapidly advancing technologies in molecular biology (that have prompted the expansion of the title of this book) on our understanding of the mechanisms of antimicrobial action and drug resistance has been remarkable. Valuable new antimicrobial drugs have emerged in all areas of infectious disease, perhaps most notably in the treatment of AIDS, where combinations of several new anti-HIV compounds have brought new hope to victims of this appalling disease.

In this new edition attention concentrates largely on the action of compounds in clinical use against micro-organisms; earlier accounts of anticancer compounds have been omitted since these now form a large subject in its own right. Rapidly expanding knowledge of the molecular genetics and biochemistry of antimicrobial drug resistance has required separate chapters on each of these topics. A separate chapter has also been devoted to drugs with biochemical activities that could not be included within the main mechanisms described in Chapters 2, 3, 4 and 5.

Drs Bob Nolan and Keith Barret-Bee, who were major contributors to the fourth edition, have since moved on to new fields of endeavour and, sadly, Alan Snow died in 1995. Although I must therefore take sole responsibility for the content of this new edition, I have been greatly helped by the incisive comments of my colleagues Drs Terry Hennessey and Wright Nichols on certain sections of the book. Over the years many helpful comments and criticisms from our readers have been invaluable in planning future editions. I hope that they will continue to let me have their views.

Finally I would like to express my thanks to Zeneca Pharmaceuticals for the provision of facilities which have made this new edition possible.

TREVOR J. FRANKLIN

Chapter one

The development of antimicrobial agents past, present and future

1.1 The social and economic importance of antimicrobial agents

Few developments in the history of medicine have had such a profound effect upon human life and society as the development of the power to control infections due to micro-organisms. In 1969 the Surgeon General of the United States stated that it was time 'to close the book on infectious diseases'. His optimism, which was shared by many, seemed justified at the time. In the fight against infectious disease several factors had combined to produce remarkable achievements. The first advances were mainly the result of improved sanitation and housing. These removed some of the worst foci of infectious disease and limited the spread of infection through vermin and insect parasites or by contaminated water and food. The earliest effective direct control of infectious diseases was achieved through vaccination and similar immunological methods which still play an important part in the control of infection today. The use of antimicrobial drugs for the control of infection is almost entirely a development of this century, and the most dramatic developments had taken place only since the 1930s. No longer was surgery the desperate gamble with human life it had been in the early nineteenth century. The perils of childbirth had been greatly lessened with the control of puerperal fever. The death of children and young adults from meningitis, tuberculosis and septicaemia, once commonplace, was, by the late 1960s, unusual in the developed world.

Unfortunately, since the heady optimism of those days we have learned to our cost that microbial pathogens still have the capacity to spring unpleasant surprises on the world. The problem of acquired bacterial resistance to drugs, recognized since the early days of penicillin use in the 1940s, has become ever more menacing. Infections caused by the tubercle bacillus and *Staphylococcus aureus*, which were once readily cured by drug therapy, are now increasingly difficult or even impossible to treat because of widespread bacterial resistance to the available drugs. Nor is resistance confined to these organisms, many other species of bacteria, as well as viruses and protozoa, are also becoming drug-resistant. The ability of micro-organisms to kill or disable the more vulnerable members of society, especially the very young and old and patients with weakened immune defences, is reported in the media almost daily. Alarming reports of lethal enteric infections, meningitis and 'flesh-eating' bacteria have become depressingly familiar. If this were not enough, the spectre of the virus (HIV) infection which leads to AIDS (acquired immune deficiency syndrome) threatens human populations around the world, in nations both rich and poor. While drug therapy for AIDS is increasingly effective, other terrifying viral infections such as Ebola and Lassa fever, for which there are no treatments, make their appearance from time to time. Throughout much of the tropical and subtropical world malaria continues to exact a dreadful toll on the health and lives of the inhabitants. Although mass movements of populations and the failure to control the insect vector, the anopheline mosquito, are major factors in the prevalence of malaria, the increasing resistance of the malarial protozoal parasite to drug treatment is perhaps the most worrying feature.

Another area of concern is the increasing incidence of serious infections caused by fungi. Thirty years ago such infections were relatively rare. More common infections like thrush and ringworm were more of an unpleasant nuisance than a serious threat to health. Today, however, many patients with impaired immunity caused by HIV infection, cytotoxic chemotherapy for malignant disease, or the immunosuppressive treatment associated with organ graft surgery, are at risk from dangerous fungal pathogens such as *Pneumocystis carinii* and *Cryptococcus neoformans*.

Fortunately, despite the threats posed by drug-resistant bacteria, viruses, protozoal parasites and fungal pathogens, the current scene is not one of unrelieved gloom. Most bacterial and fungal infections can still be treated successfully with the remarkable array of drugs available to the medical (and veterinary) professions. Work continues to develop drugs effective against resistant pathogens and significant progress can be reported against HIV and herpes infections. Vaccines have been remarkably successful in preventing some bacterial and viral infections. Indeed, outstanding amongst the medical achievements of the twentieth century have been the eradication of smallpox and the dramatic reduction in the incidence of poliomyelitis by mass vaccination programmes.

1.2 An outline of the historical development of antimicrobial agents

1.2.1 Early remedies

Among many traditional and folk remedies three sources of antimicrobial compounds have survived to the present day. These are cinchona bark and quinghaosu for the treatment of malaria and ipecacuanha root for amoebic dysentery. Cinchona bark was used by the Indians of Peru for treating malaria and was introduced into European medicine by the Spanish in the early seventeenth century. The active principle, quinine, was isolated in 1820. Quinine remained the only treatment for malaria until well into the twentieth century and still has a place in chemotherapy. The isolation of artemisinin, the active compound in the traditional Chinese remedy, quinghaosu, is much more recent and only in recent years has its therapeutic potential against malaria been fully appreciated. Ipecacuanha root was known in Brazil and probably in Asia for its curative action in diarrhoeas and dysentery. Emetine was isolated as the active constituent in 1817 and was shown in 1891 to have a specific action against amoebic dysentery. In combination with other drugs it is still used for treating this disease. These early remedies were used without any understanding of the nature of the diseases. Malaria, for example, was thought to be caused by 'bad air' (mal'aria) arising from marshy places; the significance of the blood-borne parasite was not recognized until 1880 and only in 1897 was the anopheline mosquito proved to be the specific insect vector when the developing parasite was observed in the intestine of the mosquito.

1.2.2 Antiseptics and disinfectants

The use of disinfectants and antiseptics also preceded an understanding of their action, and seems to have arisen from the observation that certain substances stopped the putrefaction of meat or rotting of wood. The term 'antiseptic' itself was apparently first used by Pringle in 1750 to describe substances that prevent putrefaction. The idea was eventually applied to the treatment of suppurating wounds. Mercuric chloride was used by Arabian physicians in the Middle Ages for preventing sepsis in open wounds. However, it was not until the nineteenth century that antiseptics came into general use in medicine. Chlorinated soda, essentially hypochlorite, was introduced in 1825 by Labarraque for the treatment of infected wounds, and tincture of iodine was first used in 1839. One of the earliest examples of disinfection used in preventing the spread of infectious disease was recorded by Oliver Wendel Holmes in 1835. He regularly washed his hands in a solution of chloride of lime when dealing with cases of puerperal fever and thereby greatly reduced the incidence of fresh infections, as did Ignaz Semmelweiss in Vienna a few years later. These pioneer attempts at antisepsis were not generally accepted until Pasteur's publication in 1863 of the microbial origin of putrefaction. This led to an understanding of the origin of infection and suggested the rationale for its prevention. As so often in the history of medicine, a change of practice depended upon the personality and persistence of one man. In antiseptics this man was Lister. He chose phenol, the antiseptic that had been introduced by Lemaire in 1860, and applied it vigorously in surgery. A 2.5% solution was used for dressing wounds and twice this concentration for sterilizing instruments. Later he used a spray of phenol solution to produce an essentially sterile environment for carrying out surgical operations. The previous state of surgery had been deplorable; wounds usually became infected and the mortality was appalling. The effect of Lister's measures was revolutionary and the antiseptic technique opened the way to great surgical advances. Even at this time, about 1870, the use of antiseptics was still empirical. An understanding of their function began with the work of Koch, who from 1881 onwards introduced the techniques on which modern bacteriology has been built. He perfected methods of obtaining pure cultures of bacteria and of growing them on solid media and he demonstrated practical methods of sterile working. Once it became possible to handle bacteria in a controlled environment the action of disinfectants and antiseptics could be studied. The pioneer work on the scientific approach to this subject was published by Kronig and Paul in 1897.

Since that time the history of antiseptics has been one of steady but unspectacular improvement. Many of the traditional antiseptics have continued in use in refined forms. The phenols have been modified and made more acceptable for general use. Acriflavine, introduced in 1913, was the first of a number of basic antiseptics. It had many years of use but was displaced by colourless cationic antiseptics (acriflavine is bright orange). In surgery the antiseptic era gave way to the aseptic era in which the emphasis is on the avoidance of bacterial contamination rather than on killing bacteria already present. All the same, infection of surgical wounds remains a constant risk and antiseptics are still used as an extra precaution or second line of defence. Surgical staff also 'scrub up' with mild antiseptic solutions before entering the operating theatre. Disinfectants play an important part in the hygiene of milking sheds, broiler houses and other places where strict asepsis is impracticable.

1.2.3 The beginnings of chemotherapy

The publications of Pasteur and Koch firmly established that micro-organisms are the cause of infectious disease, though for some diseases the causative organism still remained to be discovered. It was also known that bacteria are killed by various antiseptics and disinfectants. Not surprisingly attempts were made to kill micro-organisms within the body and so to end the infection. Koch himself carried out some experiments with this aim. He had shown that mercuric chloride is one of the few disinfectants able to kill the particularly tough spores of the anthrax bacillus. Koch therefore tried to cure animals of anthrax infection by injecting mercuric chloride. Unfortunately the animals died of mercury poisoning and their organs still contained infectious anthrax bacilli. A slightly more successful attempt to cure an infection with a toxic agent was made by Lindgard in 1893. He treated horses suffering from surra, a disease now known to be caused by trypanosomes, with arsenious oxide. There was some improvement of the disease, but the compound was too toxic to be generally useful.

However, chemotherapy really began with Paul Ehrlich. During the 10 years from 1902 onwards Ehrlich's work foreshadowed almost all the concepts which have governed subsequent work on synthetic antimicrobial agents. His first ideas arose from studies with 'vital stains', dyestuffs that were taken up selectively by living tissue. One such dye was methylene blue, which in the animal body is concentrated in nervous tissue. Ehrlich showed that the same dye was readily taken up by the malaria parasites in the blood so that they become deeply stained. Consequently methylene blue was tried against human malaria and showed some effect, though not sufficient to make it a useful treatment. Nevertheless this minor success started a line of thought that was to prove of the greatest significance. Ehrlich believed that antimicrobial agents must be essentially toxic compounds and that they must bind to the micro-organism in order to exert their action. The problem was to discover compounds having a selective action against the microbial cell compared with the cells of the host animal. Starting from methylene blue, Ehrlich began to search for other dyestuffs that would affect protozoal diseases. In 1904, after testing hundreds of available dyes, he eventually found one that was effective against trypanosomiasis in horses. This compound, called trypan red, was a significant landmark in the treatment of microbial infections since it was the first man-made compound that produced a curative effect.

However, it was not in the field of dyestuffs that Ehrlich achieved his greatest success. Following the early work on the treatment of trypanosomiasis with arsenious oxide, Koch tested the organic arsenical atoxyl (Figure 1.1). This compound produced the first cures of sleeping sickness, a human trypanosomal disease. Unfortunately, however, the compound produced serious side-effects, some patients developing optic atrophy. The curative effect of this compound stimulated Ehrlich to make other related arsenicals. He tested these on mice infected experimentally with trypanosomiasis and showed that curative action did not run parallel with toxicity to the mice. This suggested that if enough compounds were made some would have sufficiently low toxicity to be safe as chemotherapeutic agents. Ehrlich continued his search for compounds active against various micro-organisms and showed that arsenicals were active against the causative organism of syphilis. He began a massive search for an organoarsenical compound that could be used in the treatment of this disease and eventually in 1910 discovered the famous drug salvarsan (Figure 1.1). This drug and its derivative neo-salvarsan became the standard treatment for syphilis. Coupled with bismuth therapy they remained in use until supplanted by penicillin in 1945. This was the most spectacular practical achievement of Ehrlich's career, but scientifically he is remembered at least as much for his wealth of ideas that have inspired workers in the field of chemotherapy down to the present day. These ideas are so important that they deserve separate consideration.

1.2.4 The debt of chemotherapy to Ehrlich

The very term chemotherapy was invented by Ehrlich and expressed his belief that infectious disease could be combated by treatment with synthetic chemicals. Successes since his day have entirely justified his faith in this possibility. He postulated that cells possess chemical receptors which are concerned with the uptake of nutrients. Drugs that affect the cell must bind to one or other of these receptors. The toxicity of a drug is determined partly by its distribution in the body. However, in the treatment of an infection the binding to the parasite relative to the host cell determines the effectiveness of the compound.

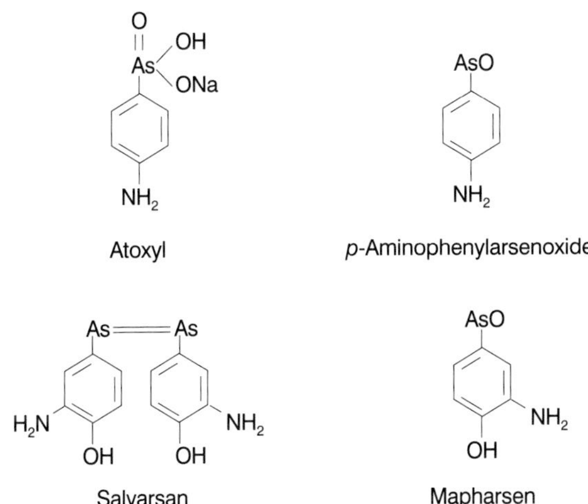

FIGURE 1.1 Arsenical compounds used in the early treatment of trypanosomiasis or syphilis.

Thus Ehrlich recognized the importance of quantitative measurement of the relationship between the dose of a compound required to produce a therapeutic effect and the dose that causes toxic reactions. Such measurements are still of prime importance in chemotherapy today.

Ehrlich pioneered methods that have since become the mainstay of the search for new drugs. One aspect of his approach was the use of screening. This is the application of a relatively simple test to large numbers of compounds in order to obtain evidence of biological activity in types of chemical structure not previously examined. The second of Ehrlich's methods was the deliberate synthesis of chemical variants of a compound known to have the required activity. The new compounds were examined for increased activity or for improvements in some other property such as reduced toxicity. Any improvement found was used as a guide to further synthesis, eventually arriving, by a series of steps, at the best possible compound. These methods are now so well accepted that their novelty in Ehrlich's day can easily be forgotten. They depend on the thesis that a useful drug possesses an ideal combination of structural features which cannot be predicted at the outset. A compound approximating to this ideal will show some degree of activity, and can therefore act as a 'lead' towards the best attainable structure.

According to Ehrlich a chemotherapeutic substance has two functional features, the 'haptophore' or binding group which enables the compound to attach itself to the cell receptors, and the 'toxophore' or toxic group that brings about an adverse effect on the cell. This idea has had a continuing influence in subsequent years. In cancer chemotherapy it has frequently been used in attempts to bring about the specific concentration of toxic agents or antimetabolites in tumour cells. In antimicrobial research it has helped to explain some features of the biochemical action of antimicrobial compounds.

Ehrlich also recognized that compounds acting on microbial infection need not necessarily kill the invading organism. It was, he suggested, sufficient to prevent substantial multiplication of the infectious agent, since the normal body defences, antibodies and phagocytes, would cope with foreign organisms provided that their numbers were not overwhelming. His views on this topic were based in part on his observation that isolated spirochaetes treated with low concentrations of salvarsan remained motile and were therefore apparently still alive. Nevertheless they were unable to produce an infection when they were injected into an animal body. It is a striking fact that several of today's important antibacterial and antifungal drugs are 'static' rather than 'cidal' in action.

Another feature of Ehrlich's work was his recognition of the possibility that drugs may be activated by metabolism in the body. This suggestion was prompted by the observation that the compound atoxyl was active against trypanosomal infections but was inactive against isolated trypanosomes. His explanation was that atoxyl was reduced in the body to the much more toxic *p*-aminophenylarsenoxide (Figure 1.1). Later work showed that atoxyl and other related arsenic acids are not in fact readily reduced to arsenoxides in the body but local reduction by the parasite remains a possibility. Ehrlich, surprisingly, did not recognize that his own compound salvarsan would undergo metabolic cleavage. In animals it gives rise to the arsenoxide as the first of a series of metabolites. This compound eventually was introduced into medicine in 1932 under the name mapharsen (Figure 1.1); its toxicity is rather high, but it has sufficient selectivity to give it useful chemotherapeutic properties. Other examples of activation through metabolism have been discovered in more recent times, for example the conversion of the antimalarial 'prodrug' proguanil to the active cycloguanil in the liver and the metabolism of antiviral nucleosides to the inhibitory triphosphate derivatives.

Ehrlich also drew attention to the problem of resistance of micro-organisms towards chemotherapeutic compounds. He noticed it in the treatment of trypanosomes with parafuchsin and later with trypan red and atoxyl. He found that resistance extended to other compounds chemically related to the original three, but there was no cross-resistance between the groups. In Ehrlich's view this was evidence that each of these compounds was affecting a separate receptor. Independent resistance to different drugs later became a commonplace in antimicrobial therapy. Ehrlich's view of the nature of resistance is also interesting. He found that trypanosomes resistant to trypan red absorbed less of the compound than sensitive strains, and he postulated that the receptors in resistant cells had a diminished affinity for the dye. This mechanism corresponds to one of the currently accepted types of resistance in micro-organisms (Chapter 9).

Several useful antimicrobial drugs arose in later years as an extension of Ehrlich's work. The most notable (Figure 1.2) are suramin, a development from trypan red, and mepacrine (also known as quinacrine or atebrin) indirectly from methylene blue (Figure 1.2). Suramin, introduced in 1920, is a colourless compound having a useful action against human trypanosomiasis. Its particular value lay in its relative safety compared with other antimicrobial drugs of the period. It was the first useful antimicrobial drug without a toxic metal atom and the ratio of the dose required to produce toxic symptoms to that needed for a curative effect is vastly higher than with any of the arsenicals. Suramin is remarkably persistent, a single dose giving protection for more than a month. Mepacrine, first marketed in 1933, was an antimalarial agent of immense value in the Second World War. It was supplanted by other compounds partly because it caused a yellow discoloration of the skin. Besides these obvious descendants from Ehrlich's work the whole field of drug therapy is permeated by his ideas and many other important compounds can be traced directly or indirectly to the influence of his thought.

1.2.5 The treatment of bacterial infections by synthetic compounds

In spite of the successes achieved in the treatment of protozoal diseases and the spirochaetal disease syphilis, the therapy of bacterial infections remained for many years an elusive and apparently unattainable goal. Ehrlich himself, in collaboration with Bechtold, made a series of phenols which showed much higher antibacterial potency than the simple phenols originally used as disinfectants. However, these compounds had no effect on bacterial infections in animals. Other attempts were equally unsuccessful and no practical progress was made until 1935, when Domagk reported the activity of prontosil rubrum (Figure 4.1) against infections in animals. The discovery occurred in the course of a widespread research programme on the therapeutic use of dyestuffs, apparently inspired by Ehrlich's ideas. Trefouel showed that prontosil rubrum is broken down in the body giving sulphanilamide (Figure 4.1) which was in fact the effective antibacterial agent. The sulphonamides were outstandingly successful drugs. They might have been developed and used even more widely if penicillin and other antibiotics had not followed on so speedily. Surprisingly in the field of synthetic antibacterial agents, relatively few other successes have been achieved against the common bacterial infections. In part this may have

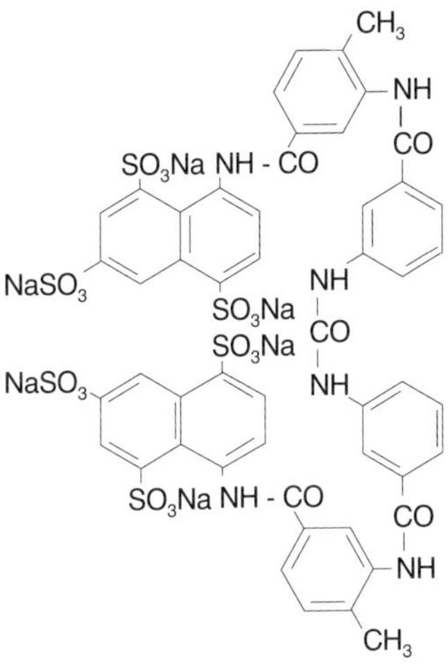

Suramin

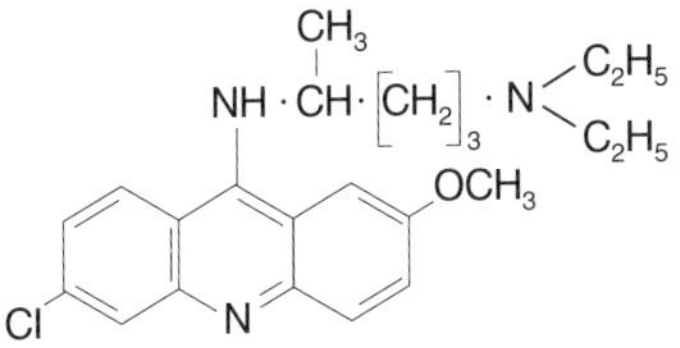

Mepacrine

FIGURE 1.2 Early synthetic compounds used for treating protozoal infections: suramin for trypanosomiasis (sleeping sickness) and mepacrine for malaria.

been due to the diminished incentive to search for new products when the antibiotics proved to give such excellent results. However, synthetic compounds with a useful action against bacterial infections have certainly been difficult to find. Other synthetic compounds in current use include the quinolones, trimethoprim, certain imidazoles such as metronidazole, nitrofurans such as nitrofurantoin, and several drugs used in the treatment of tuberculosis (see below). It should also be remembered that modern β-lactam drugs (Chapter 2) are semi-synthetic variants of naturally occurring molecules. For several years after treatment was available for streptococcal and staphylococcal infections, the mycobacterial infections that cause tuberculosis and leprosy remained untreatable by chemotherapy. The first success came with the antibiotic streptomycin, which remains an optional part of the standard treatment for tuberculosis. Later several chemically unrelated synthetic agents were also found to be effective against this disease. The best of these are isonicotinyl hydrazide (isoniazid), ethambutol and pyrazinamide. An antibiotic, rifampicin (rifampin) is usually included in the combination therapy for tuberculosis. In leprosy the drug regularly used is also a synthetic compound, 4,4'-diaminodiphenylsulphone.

1.2.6 The antibiotic revolution

Ever since bacteria have been cultivated on solid media, contaminant organisms have occasionally appeared on the plate. Sometimes this foreign colony would be surrounded by an area in which bacterial growth was suppressed. Usually this was regarded as a mere technical nuisance, but Alexander Fleming in 1928, observing such an effect with a mould *Penicillium notatum* on a plate seeded with staphylococci, was struck by its potential importance. He showed that the mould produced a freely diffusible substance that was highly active against Gram-positive bacteria, and apparently of low toxicity to animals. He named it penicillin. It was, however, unstable and early attempts to extract it failed, so Fleming's observation lay neglected until 1939. By then the success of the sulphonamides had stimulated a renewed interest in the chemotherapy of bacterial infections. The search for other antibacterial agents now seemed a promising and exciting project and Florey and Chain selected Fleming's penicillin for re-examination. They succeeded in isolating an impure but highly active solid preparation and published their results in 1940. Evidence of its great clinical usefulness in humans followed in 1941. It was now apparent that a compound of revolutionary importance in medicine had been discovered. To make it generally available for medical use, however, presented formidable problems both in research and in large-scale production, especially under conditions of wartime stringency. Eventually perhaps the biggest chemical and biological joint research programme ever mounted was undertaken, involving 39 laboratories in Britain and the United States. It was an untidy operation, with much duplication and overlapping of work, but it culminated in the isolation of pure penicillin, the determination of its structure and the establishment of the method for its production on a large scale. The obstacles overcome in this research were enormous. They arose mainly from the very low concentrations of penicillin in the original mould cultures and from the marked chemical instability of the product. In the course of this work the concentration of penicillin in mould culture fluids was increased 1000-fold by the isolation of improved variants of *Penicillium notatum*, using selection and mutation methods and by improved conditions of culture. This tremendous improvement in yield was decisive in making large-scale production practicable and ultimately cheap.

The success of penicillin quickly diverted a great deal of scientific effort towards the search for other antibiotics. The most prominent name in this development was that of Waksman, who began an intensive search for antibiotics in micro-organisms isolated from soil samples obtained in many parts of the world. Waksman's first success was streptomycin and many other antibiotics followed. His screening methods were copied in many other laboratories. Organisms of all kinds were examined and hundreds of thousands of cultures were tested. Further successes came quickly. Out of all this research several thousand named antibiotics have been listed. However, most of them have defects that prevent their development as drugs. Perhaps 50 have had some sort of clinical use and very few of these are regularly employed in the therapy of infectious disease. However, among this select group and their semi-synthetic variants are compounds of such excellent qualities that treatment is now available for most of the bacterial infections known to occur in humans, although, as we have seen, resistance increasingly threatens the efficacy of drug therapy.

Following the wave of discovery of novel classes of antibiotics in the 1940s and 1950s, research focused largely on taking antibiotics of proven worth and subjecting them to chemical modification in order to extend their antibacterial spectrum, to combat resistance and to improve their acceptability to patients. Recently, however, the pressure of increasing drug resistance has led to renewed attention to the discovery of novel chemical classes of both naturally occurring and synthetic antibacterial compounds

1.2.7 Antifungal and antiviral drugs

The diversity of fungal pathogens which attack humans and domesticated animals is considerably smaller than that of bacteria. Nevertheless, fungi cause infections ranging from the trivial and inconvenient to those causing major illness and death. Fungal infections have assumed greater importance in recent years because of the increased number of medical conditions in which host immunity is compromised. Fungi, as eukaryotes, have much more biochemistry in common with mammalian cells than bacteria and therefore pose a serious challenge to chemotherapy. Specificity of action is more difficult to achieve. Few antibiotics are useful against fungal infections and attention has concentrated more on devising synthetic agents. Some advantage has been taken of the progress in producing compounds for the treatment of fungal infections of plants, in order to develop from them reasonably safe and effective drugs against human fungal infections.

Enormous strides have been made in the control of viral infections through the use of vaccines. As mentioned previously, smallpox has been eradicated throughout the world. In the developed countries at least the seasonal epidemics of poliomyelitis, which were the cause of so much fear and suffering 40 years ago, have disappeared. But despite these and other vaccine-led successes against viral infections, not all such infections can be controlled so effectively by mass vaccination programmes. The bewildering diversity of common cold viruses, the ever-shifting antigenic profiles of influenza viruses and the insidious nature of the virus that leads to AIDS are just three examples of diseases that may not yield readily to the vaccine approach. Attention is therefore focused on finding drugs that specifically arrest viral replication, a formidable challenge since viruses partially parasitize the biochemistry of the host cells. The naturally occurring antiviral protein (interferon-α (IFNα), Chapter 6), now readily available as a recombinant protein, has a useful role in the treatment of the dangerous hepatitis B and C viruses.

1.2.8 Antiprotozoal drugs

After the Second World War several valuable new drugs were introduced into the fight against malaria, including chloroquine, proguanil and pyrimethamine. For several years these drugs were extremely effective both for the prevention and treatment of malaria. However, by the time of the outbreak of the war in Vietnam in the 1960s it had become clear that, like bacteria, the malarial parasites were adept at finding ways to resist drug therapy. The US government then launched a massive screening project to discover new antimalarial agents. Two compounds, mefloquine and halofantrine, resulted from this effort and are still in use today. Nevertheless, the development of resistance to these drugs seems inevitable and the search for new antimalarial drugs continues. Two compounds currently appear to offer real promise: artemisinin, a naturally occurring compound isolated from qinghaosu, and the synthetic compound atovaquone (Chapter 6).

The treatment of other serious protozoal infections, such as the African and South American forms of trypanosomiasis, remains relatively primitive. The arsenical melarsoprol is still used for African trypanosomiasis (sleeping sickness), although the less toxic difluorodimethyl ornithine is increasingly seen as the drug of choice. South American trypanosomiasis, or Chagas disease, is still very difficult to treat successfully, and control of the insect vector, the so-called 'kissing bug' which infests poor-quality housing, is very effective. The only useful drugs against leishmaniasis are such venerable compounds as sodium antimony gluconate and pentamidine, neither of which is ideal. Unfortunately, the parasitic diseases of the developing world do not present the major pharmaceutical companies with attractive commercial opportunities, and research into the treatment of these diseases is relatively neglected.

1.3 Reasons for studying the biochemistry and molecular biology of antimicrobial compounds

Following this brief survey of the discovery of the present wide range of antimicrobial compounds, we may now turn to the main theme of the book. We shall be concerned with the biochemical mechanisms that underlie the action of compounds used in the battle against pathogenic micro-organisms. This topic has a twofold interest. In the long run, a detailed understanding of drug action at the molecular level may generate ideas for the design of entirely novel antimicrobial agents. In spite of the great power and success of the drugs currently available, there are still areas where improvements could be made. The other interest in antimicrobial agents is the light that their activity can throw on the subject of biochemistry itself. Antibacterial agents, particularly the antibiotics, often have a highly selective action on biochemical processes. They may block a single reaction within a complex sequence of events. The use of such agents has often revealed details of biochemical processes that would otherwise have been difficult to disentangle. Understanding of the biochemistry of antimicrobial action has been built up slowly and painfully, with many false starts and setbacks. Since about 1960, however, a much greater insight into the action of antibacterial compounds has been achieved, and for most of the compounds commonly used in medical practice at least an outline can be given of the biochemical effects underlying their antimicrobial action. Knowledge of the mechanism of action of antiprotozoal drugs, some of which were discovered long before the antibacterial drugs, lagged well behind for many years. This was due mainly to the difficulty in isolating and working with protozoa outside the animal body, but interest had also been concentrated on bacteria because of their special importance in infectious disease and because of their widespread use in biochemical and genetic research. However, advances in the molecular genetics of the major parasitic protozoans should now greatly facilitate the development of our understanding of drug action in these species. Rapid progress is also being made in working out the biochemical and molecular basis of the action of antifungal and antiviral drugs.

1.4 Uncovering the molecular basis of antimicrobial action

An understanding of how antimicrobial compounds work has developed gradually. Several levels of progress can be distinguished and will be discussed separately.

1.4.1 Nature of the biochemical systems affected

As long as antimicrobial compounds have been known, workers have attempted to explain their action in biochemical terms. Ehrlich made a tentative beginning in this direction when he suggested that the arsenicals might act by combining with thiol groups on the protozoal cells. However, he was severely limited by the elementary state of biochemistry at that time. By the time the sulphonamides were discovered, the biochemistry of small molecules was much more advanced and reasonable biochemical explanations of sulphonamide action were soon available. However, many of the antibiotics that followed presented very different problems. Attempts to apply biochemical methods to the study of their action led to highly conflicting answers. At one stage a count showed that 14 different biochemical systems had been suggested as the site of action of streptomycin against bacteria. Much of this confusion arose from a failure to distinguish between primary and secondary effects. The biochemical processes of bacterial cells are closely interlinked. Hence disturbance of any one important system is likely to have effects on many of the others. Methods had to be developed that would distinguish between the primary biochemical effect of an antimicrobial agent and other changes in metabolism that followed as a consequence. Once these were established, more accurate assessments could be made of the real site of action of various antimicrobial compounds. The limiting factor then became the extent of biochemical information about the nature of the target site. From about 1955 onwards there has been a dramatic increase in the understanding of the structure, function and synthesis of macromolecules. Most of the important antibiotics were found to act by interfering with the biosynthesis or function of macromolecules, and the development of new techniques has provided the means of locating their site of action with some confidence.

1.4.2 Methods used for the study of the mode of action of antimicrobial compounds

Experience gained over the past four decades has helped to evolve reasonably systematic procedures for working out the primary site of action of many antimicrobial compounds. Once the primary site of action is established, the overall effect of a drug on the metabolism of microbial cells can often be explained. Many of the techniques used in elucidating the mode of action of antimicrobial agents are discussed in later chapters, but it may be helpful to set them out in a logical sequence.

1. Once the chemical structure of the drug is established, it is studied carefully to determine whether a structural analogy exists with part, or the whole, of a biologically important molecule; for example, a metabolic intermediate or

essential cofactor, nutrient, etc. An analogy may be immediately obvious, but sometimes it becomes apparent only through imaginative molecular model building or by hindsight after the target site of the compound has been revealed by other means. Analogies of structure can sometimes be misleading and should only be used as a preliminary indication.

2. The next step is to examine the effects of the compound on the growth kinetics and morphology of suitable target cells. A cytocidal effect shown by reduction in viable count may indicate damage to the cell membrane. This can be confirmed by observation of leakage of potassium ions, nucleotides or amino acids from the cells. Severe damage leads to cell lysis. Examination of cells by electron microscopy may show morphological changes which indicate interference with the synthesis of one of the components of the cell wall. Many antibiotics have only a cytostatic action and do not cause any detectable morphological changes.
3. Usually, attempts are made to reverse the action of an inhibitor by the addition to the medium of various supplements. Nutrients, including oxidizable carbon sources, fatty acids and amino acids, intermediary metabolites such as purines and pyrimidines, vitamins and bacterial growth are tested in turn. If reversal is achieved, this may point to the reaction or reaction sequence which is blocked by the inhibitor. Valuable confirmatory evidence can often be obtained by the use of a genetically engineered auxotrophic organism which requires a compound known to be the next intermediate in a biosynthetic sequence beyond the reaction blocked by the antimicrobial agent. An auxotroph of this type should be resistant to the action of the inhibitor. Inhibition in a biosynthetic sequence may also be revealed by accumulation of the metabolite immediately *before* the blocked reaction. Unfortunately, the actions of many antimicrobial agents are not reversed by exogenous compounds. This especially applies to compounds which interfere with the polymerization stages in nucleic acid and protein biosynthesis, where reversal is impossible.
4. The ability of an inhibitor to interfere with the supply and consumption of ATP is usually examined since any disturbance to energy metabolism has profound effects on the biological activity of the cell. The inhibitor is tested against the respiratory and glycolytic activities of the micro-organism, and the ATP content of the cells is measured.
5. Useful information can often be gained by observing the effect of an antimicrobial agent on the kinetics of uptake of a radiolabelled nutrient, such as glucose, acetate, a fatty acid, an amino acid, phosphate, etc. Changes in rate of incorporation which follow the addition of the drug are measured and compared with its effect on growth. A prompt interference with incorporation of a particular nutrient may provide a good clue to the primary site of action.
6. An antimicrobial compound which inhibits protein or nucleic acid synthesis in cells without interfering with membrane function or the biosynthesis of the immediate precursors of proteins and nucleic acids or the generation and utilization of ATP, probably inhibits macromolecular synthesis directly. Because of the close interrelationship between protein and nucleic acid synthesis, indirect effects of the inhibition of one process on the other must be carefully distinguished. For example, inhibitors of the biosynthesis of RNA also block protein biosynthesis as the supply of messenger RNA is exhausted. Again, inhibitors of protein synthesis eventually arrest DNA synthesis because of the requirement for continued protein biosynthesis for the initiation of new cycles of DNA replication. A study of the kinetics of the inhibition of each macromolecular biosynthesis in intact cells is valuable since indirect inhibitions appear later than direct effects.
7. When the inhibited biochemical system has been identified in intact cells, more detailed information can then be obtained with prepa-

rations of enzymes, nucleic acids and subcellular organelles. The antimicrobial compound is tested for inhibitory activity against the suspected target reaction *in vitro*. There is a danger, however, of non-specific drug effects *in vitro*, especially at high concentrations. On the other hand, failure to inhibit the suspected target reaction *in vitro*, even with very high concentrations of drug, cannot rule out inhibition of the same reaction in intact cells, for several reasons:

(a) the drug may be metabolized by the host or by the micro-organism to the active, inhibitory derivative;
(b) the procedures involved in the purification of an enzyme may desensitize it to the inhibitor by altering the conformation of the inhibitor binding site; or
(c) the site of inhibition in the intact cell may be part of a highly integrated structure which is disrupted during the preparation of a cell-free system, again causing a loss of sensitivity to the inhibitor.

The use of cell-free preparations from drug-resistant mutants is often useful in identifying the site of attack. This approach was ingeniously exploited in identifying a target site of streptomycin in bacterial ribosomes (Chapter 4). The advent of cloning and recombinant DNA technology now greatly facilitates the provision of suspected protein targets for *in vitro* evaluation.

8. The increasingly rapid acquisition of microbial genomic sequences and remarkable technological developments provide novel opportunities for profiling the effects of antimicrobial drugs by investigating their effects on the expression of hundreds or even thousands of genes. Although an antimicrobial drug may target a specific molecular receptor, its consequent effects on microbial metabolism and gene expression may not only be pleiotropic, i.e. multiple, but also characteristic. The ability to assess simultaneously the impact of drugs on the expression of many different genes enables investigators to place compounds with closely similar primary sites of action into related sets. Gene transcription profiling of novel agents with unknown sites of action may therefore provide valuable clues as to their primary target receptors.

1.4.3 The molecular interactions between antimicrobial agents and their target sites

Early mode of action studies usually concentrated on revealing the biochemical basis for the antimicrobial effect. Increasingly, scientists are no longer satisfied with this level of explanation alone and now aspire to define drug action in molecular terms, i.e. the details of the specific interactions between the drug and its target site. In order to achieve this level of understanding of drug action, specialized techniques such as X-ray crystallography and nuclear magnetic resonance spectroscopy (NMR) are used to reveal the detailed three-dimensional structures of drugs and their interacting proteins or nucleic acids. Recombinant DNA technology provides the means to identify the role of specific amino acids or nucleotide residues in macromolecule–drug interactions. However, the structural elucidation of supramolecular organized structures such as membranes and ribosomes has proved to be a more formidable undertaking. Nevertheless, the combination of advanced physical techniques and recombinant DNA methodologies provides analytical tools of unprecedented power to probe drug interactions, even at this level. The study of the precise structural requirements of drug molecules for antimicrobial activity has provided rich returns. The structures of the more complex antibiotics are often highly specific for biological activity and even minor chemical changes may result in complete inactivation. An antimicrobial agent suitable for systemic use will have a combination of properties: good absorption from the site of administration, appropriate distribution within the body, adequate persistence in the tissues, effective penetration into the micro-organism and selective inhibition of the microbial target site. Each of these attributes may require distinct molecular characteristics. For optimum activity all these characteristics must be combined in the same molecule.

1.4.4 Pharmacological biochemistry

When an antimicrobial agent is used systemically its effectiveness is determined by various factors that govern its behaviour in the body. The absorption, distribution, metabolism and excretion of drugs are essential subjects for investigation. Activity requires the maintenance of an appropriate concentration of the drug at the site of action; this concentration must be sustained for long enough to allow the body's defences time to defeat the infection. The concentration depends on relative rates of absorption and excretion and also on any metabolism of the drug. In most cases, metabolism by the host inactivates the drug, but several examples are known where metabolism is essential for converting an inactive administered compound, or prodrug, into the active entity. The degree of binding of the drug to host proteins may also be important. For example, some drugs bind strongly to plasma proteins. While this can increase their persistence in the body it can also lower their effectiveness, since drug activity depends on the concentration of free (unbound) compound at the target site and, in the case of extensively protein-bound compounds, the concentration of free drug is low. The methods for studying such factors, using modern analytical techniques, are well established. The data can help to explain species differences in the therapeutic activities of drugs and provide a sound basis for recommendations on the size and frequency of doses for treating patients.

1.4.5 Selectivity of action of antimicrobial agents

A successful antimicrobial drug must be selective in its action, and reasons have been sought for this selectivity. The basis of selectivity varies from one drug to another. For example, in some cases the biological process inhibited by a drug is specific to the microbial cell, whereas other drugs act on biochemical mechanisms common to both microbe and host. In the latter situation the basis of selectivity will need further study. Possible explanations could include:

1. significant differences between the microbial and host enzymes catalysing the same reaction that allows a selective attack on the microbial enzyme; and
2. specificity arising from selective concentration of the drug within the microbial cell.

In the latter situation the question shifts to the basis of the selective concentration. Differences between the rates of turnover of target molecules in the pathogen and its host can also provide a basis for selectivity of action (see eflornithine, Chapter 6).

1.4.6 The biochemistry of microbial resistance

As we have seen, the therapeutic value of antimicrobial drugs usually declines after sustained use due to the emergence of drug-resistant organisms. This enormously important problem has been studied in great depth by microbiological, biochemical and molecular genetic methods. Such studies reveal the genetic basis for the emergence of resistant organisms and also define the biochemical mechanisms of resistance. The mechanisms of some forms of resistance, especially in non-bacterial pathogens, still need further study, and there is obviously great practical interest in limiting the development of resistance and in finding ways of overcoming it following its inevitable appearance.

1.5 Scope and layout of the present book

In this book we have tried to select well-established evidence for the biochemical action of many of the best-known agents used in medicine.

Although much of the content is devoted to antibacterial drugs, which constitute by far the largest group of antimicrobial agents in use today, we have also brought together information on the biochemical activities of examples of commonly used antifungal, antiprotozoal and antiviral drugs. One chapter is devoted to the means by which antimicrobial compounds enter and leave their target cells. The last two chapters consider respectively the genetic and biochemical basis of drug resistance.

Wherever possible, we have grouped drugs according to their types of biochemical action rather than by their therapeutic targets. However, one chapter brings together several drugs with other and unusual modes of action.

Further reading

Garrett, L. (1995). *The Coming Plague*. Virago Press, London

Greenwood, D. (1995). *Antimicrobial Chemotherapy*, 3rd edn, Oxford University Press, Oxford.

Sneader, W. (1985). *Drug Discovery: The Evolution of Modern Medicines*, John Wiley & Sons.

Tenover, F. C. and Hughes, J. M. (1996). The challenge of emerging infectious diseases. *J. Am. Med. Assoc.* 275, 300.

Chapter two

Vulnerable shields – the cell walls of bacteria and fungi

2.1 Functions of the cell wall

In the search for differences between microbial pathogens and animal cells that could provide the basis for selective antimicrobial attack, one evident distinction lies in their general structure. The animal cell is relatively large and has a complex organization; its biochemical processes are compartmentalized and different functions are served by the nucleus with its surrounding membrane, by the mitochondria and by various other organelles. The cytoplasmic membrane is thin and lacks rigidity. The cell exists in an environment controlled in temperature in mammals and birds and also in osmolarity. It is constantly supplied with nutrients from the extracellular fluid. Bacteria and fungi live in variable and often hostile environments and they must be able to withstand considerable changes in external osmolarity. Some micro-organisms have relatively high concentrations of low molecular weight solutes in their cytoplasm. Such cells suspended in water or in dilute solutions develop a high internal osmotic pressure. This would inevitably disrupt the cytoplasmic membrane unless it were provided with a tough, elastic outer coat. This coat is the cell wall, a characteristic of bacteria and fungi which is entirely lacking in animal cells. It has a protective function but at the same time it is vulnerable to attack, and a number of antibacterial and antifungal drugs owe their action to their ability to disturb the processes by which the walls are synthesized. Since there is no parallel biosynthetic mechanism in animal cells, substances affecting this process may be highly selective in their antimicrobial action.

The term 'wall' will be used to describe all the cell covering which lies outside the cytoplasmic membrane. The structures of the walls of bacteria and fungi are very different from each other, as are the biosynthetic processes involved in their elaboration. This results in susceptibility to quite distinct antimicrobial agents.

2.2 Structure of the bacterial wall

The structure of the bacterial wall not only differs markedly from that of fungi but also varies considerably from one species to another. It nevertheless follows general patterns which are related to the broad morphological classification of bacteria. Classically this has been based on the responses towards the Gram stain, but the well-tried division into Gram-positive and Gram-negative types has a significance far beyond that of an empirical staining reaction. The most evident differences are worth recalling.

Many Gram-negative bacteria are highly adaptable organisms which can use inorganic nitrogen compounds, mineral salts and a simple carbon source for the synthesis of their whole structure. Their cytoplasm has a relatively low osmolarity. Typical Gram-positive cocci or bacilli tend to be more exacting in their nutritional needs. They are usually cultivated on rich undefined broths or on fairly elaborate synthetic media. In their cytoplasm, Gram-positive bacteria concentrate amino acids, nucleotides and other metabolites of low molecular weight and consequently have a high internal osmolarity. However, not all bacteria fit this neat division. The Gram-negative cocci, the rickettsias, the chlamydias and the spirochaetes, for example, are all Gram-negative bacteria with exacting growth requirements. The mycoplasmas lack a rigid wall structure and although technically Gram-negative, they are best treated as a separate group lying outside the usual Gram-stain classification.

For many years the bacterial wall was considered to be a rigid structure, largely because when bacteria are disrupted the isolated walls retain the shape of the intact organisms. More recent evidence, however, shows that this concept of rigidity must be revised. The peptidoglycan sacculus of the bacterial wall (see below) can expand or contract in response to changes in the ionic strength or the pH of the external environment. This responsive flexibility is a property of the wall itself and can even be seen by eye when salt solutions are added to quantities of walls pelleted by centrifugation. When intact bacteria are subjected to osmotic stress, water moves through the wall and membrane into the cytoplasm. The consequent swelling of the cell, bounded by the membrane, is accommodated to some extent by the limited elasticity of the wall, although even stretchable structures break when sufficiently stressed. The wall breaks and the cell then bursts due to the turgor pressure on the thin cytoplasmic membrane.

Most of the work on wall structure has been done with Gram-positive cocci and bacilli and with enteric bacteria and other Gram-negative rods. The extent to which the structural generalizations apply to groups outside these classes is uncertain.

2.2.1 The Gram-positive wall

Many Gram-positive bacteria have relatively simple walls (Figure 2.1). The wall, which lies outside the cytoplasmic membrane, is usually between 15 and 50 nm thick. Bacteria can be broken by shaking with small glass beads and the walls separated from cytoplasmic material by washing and differential centrifugation. In electron micrographs these wall preparations resemble empty envelopes, torn in places where the cytoplasmic contents were released. The major part of the Gram-positive wall is a large polymer comprising two covalently linked components. One of these components, forming at least 50% of the wall mass, is peptidoglycan (sometimes referred to as murein or mucopeptide). Its cross-linked structure provides a tough, fibrous fabric giving strength and shape to the cell and enabling it to withstand a high internal osmotic pressure. The amount of peptidoglycan in the wall shows that it covers the cell in a multilayered structure, with cross-linking both within and between the layers. Attached to the peptidoglycan is an acidic polymer, accounting for 30–40% of the wall mass, which differs from species to species. Often this is a teichoic acid – a substituted poly(D-ribitol 5-phosphate) (see Figure 2.8) – or a substituted glycerol 3-phosphate (lipoteichoic acid). In some bacteria teichoic acid is replaced by poly(*N*-acetylglucosamine 1-phosphate) or teichuronic acid (a polymer containing uronic acid and *N*-acetylhexosamine units). Bacteria that normally incorporate teichoic acid in their walls can switch to teichuronic acid under conditions of phosphate limitation. The acidic character of the polymer attached to the peptidoglycan ensures that the cell surface is strongly polar and carries a negative charge. This may influence the passage of ions, particularly Mg^{2+} and possible ionized drugs, into the cell. The teichoic acid or other acidic polymer is readily solubilized and released from the insoluble peptidoglycan by hydrolysis in cold acid or alkali. The nature of the linkage is described later.

Other components of the Gram-positive wall vary widely from species to species. Protein is often present to the extent of 5–10%, and protein A of *Staphylococcus aureus* is apparently linked covalently to peptidoglycan. Proteins and polysaccharides frequently occur in the outermost layers

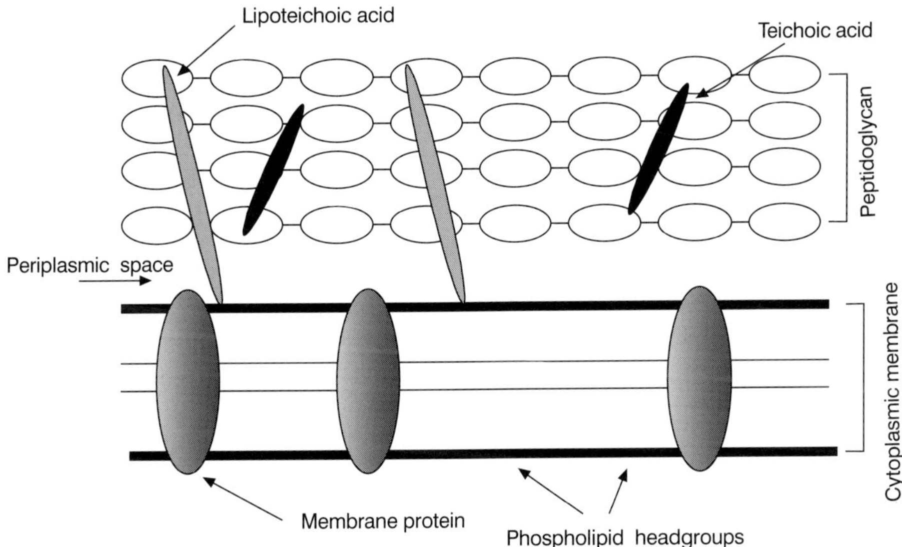

FIGURE 2.1 The arrangement of the cell envelope of Gram-positive bacteria. The components are not drawn to scale.

and provide the main source of the antigenic properties of these bacteria. Mycobacteria and a few related genera differ from other Gram-positive bacteria in having large amounts of complex lipids in their wall structure.

2.2.2 The Gram-negative wall

The Gram-negative wall is far more complex. Wide-ranging studies of its structure have been concentrated on the Enterobacteriaceae and on *Escherichia coli* in particular.

When cells of *Escherichia coli* are fixed, stained with suitable metal salts, sectioned and examined by electron microscopy, the cytoplasmic membrane is readily identified by its 'sandwich' appearance of two electron-dense layers separated by a lighter space. External to this, the cell wall appears as a structure containing three electron-dense layers separated by clear layers (Figure 2.2). The clear layer immediately outside the cytoplasmic membrane is described as the periplasmic space. Here are found soluble enzymes and other components which can be released by submitting the cell to transient non-lethal osmotic shock. The electron-dense layer, about 2 nm thick, immediately outside the periplasmic space represents the peptidoglycan component of the wall. It is much thinner than in Gram-positive bacteria and may constitute only 5–10% of the wall mass. Even so, it contributes substantially to wall strength. Cells rapidly lyse when treated with lysozyme, an enzyme which specifically degrades peptidoglycan. In *Escherichia coli* the peptidoglycan is covalently linked to a lipoprotein which probably projects into the outer regions of the wall. The outer regions of the Gram-negative cell wall have been the most difficult to characterize. The various components together form a structure 6–10 nm thick, called the outer membrane. Like the cytoplasmic membrane, it is basically a lipid bilayer (giving rise to the two outermost electron-dense bands), hydrophobic in the interior with hydrophilic groups at the outer surfaces. It also has protein components which penetrate the layer partly or completely and form the membrane 'mosaic'. Despite these broad structural similarities, the outer membrane differs widely in composition and function from the cytoplasmic membrane. Its main constituents are lipopolysaccharide, phospholipids, fatty acids and proteins. The phospholipids, mainly phosphatidylethanolamine and phosphatidylglycerol, resemble those in the cytoplasmic membrane. The structure of the lipopolysaccharide is complex and varies considerably from one bacterial strain to another. The molecule has three parts (Figure 2.3). The core is built from 3-deoxy-D-*manno*-octulosonic acid (KDO), hexoses, heptoses, ethanolamine and phosphoric acid as structural components. The three KDO residues contribute a structural unit which strongly binds the divalent ions of magnesium and calcium, an important feature stabilizing the membrane. Removal of these ions by chelating agents leads to release of some of the lipopolysaccharide into the medium; at the same time the membrane becomes permeable to compounds that would otherwise be excluded. The core polysaccharide is linked to the antigenic side chain, a polysaccharide which can vary greatly from one strain to another even within the same bacterial species. Usually it comprises about 30 sugar units, although these can vary both in number and in structure. It forms the outermost layer of the cell and is the main source of its antigenic characteristics. At the opposite end, the core of the lipopolysaccharide is attached to a moiety known as lipid A which can be hydrolysed to glucosamine, long-chain fatty acids, phosphate and ethanolamine. The fatty acid chains of lipid A, along with those of the phospholipids, align themselves to form the hydrophobic interior of the membrane. The outer membrane is therefore asymmetric, with lipopolysaccharide exclusively on the outer surface and phospholipid mainly on the inner surface.

The most abundant proteins of the outer membrane in *Escherichia coli* are the porin proteins and lipoprotein. Electron microscopy of spheroplasts lacking peptidoglycan reveals triplets of indentations in the membrane surface, each 2 nm in diam-

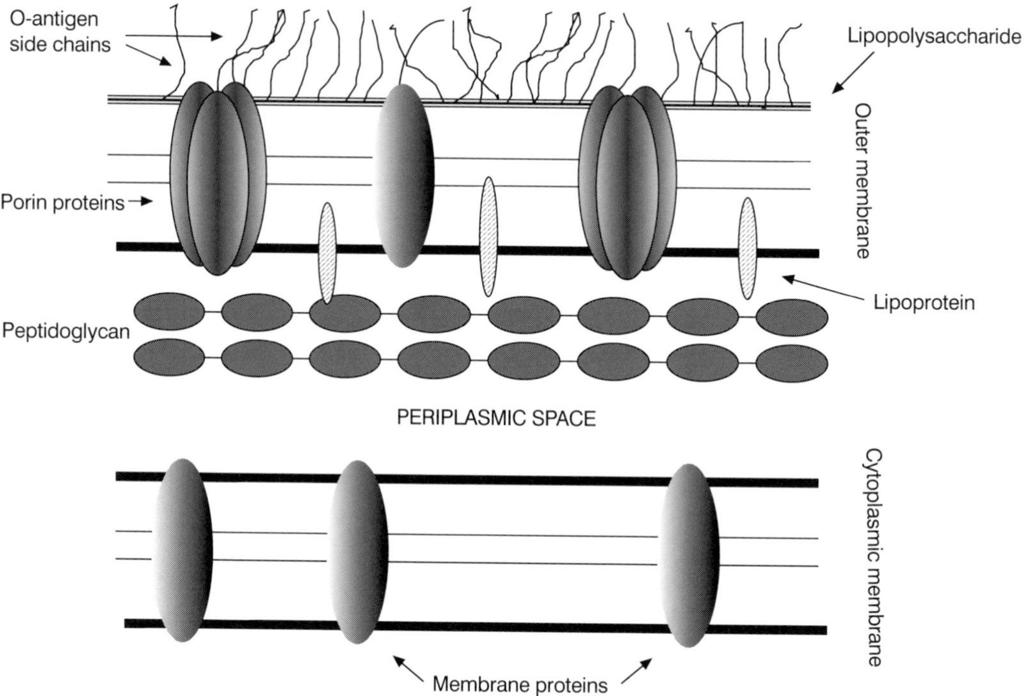

FIGURE 2.2 The arrangement of the various layers of the cell envelope of Gram-negative bacteria. The components are not drawn to scale.

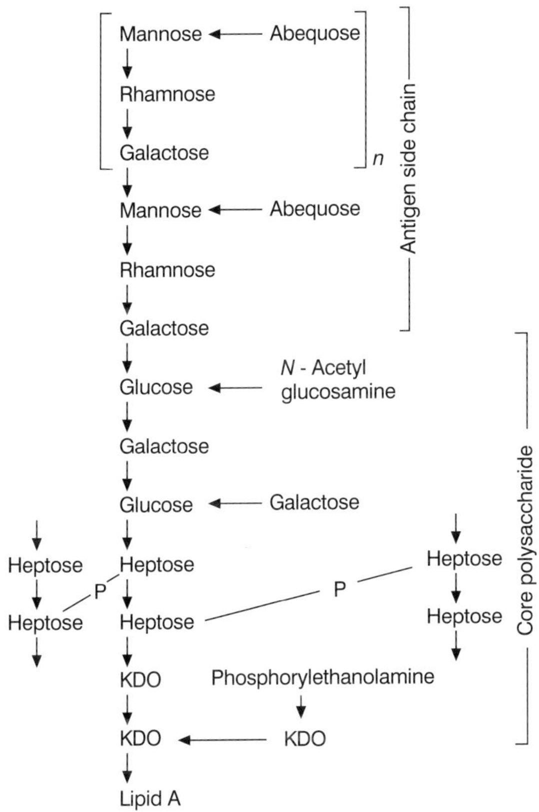

FIGURE 2.3 Structure of the lipopolysaccharide of the cell envelope of *Salmonella typhimurium*. The diagram has been simplified by omitting the configuration of the glycosidic linkages and omitting the O-acetyl groups from the abequose units. KDO: 3-deoxy-D-*manno*-octulosonic acid. Lipid A consists of a β-1,6-linked diglucosamine residue to which lauric, myristic, palmitic and 3-D(-)-hydroxymyristic acids are bound. The heptose residues of three lipopolysaccharide polymers are shown linked by phosphate diester bridges. Although there are considerable structural variations in the antigen side chains among *Salmonella* species, the core polysaccharide and lipid A are probably common to all wild-type salmonellae. The core structure in *Escherichia coli* is more variable.

eter and 3 nm apart, through which the stain used in the preparation readily penetrates. This is interpreted as showing that the porin protein molecules stretch across the membrane in groups of three, enclosing pores through which water and small molecules can diffuse. The size of the pores explains the selective permeability of the Gram-negative outer membrane, freely allowing the passage of hydrophilic molecules up to a maximum molecular weight of 600–700. Larger flexible molecules may also diffuse through the pores although with more difficulty. Artificial vesicles can be made with outer membrane lipids. Without protein these are impermeable to solutes, but when porins are incorporated they show permeability characteristics similar to those of the outer membrane itself. The role of porins in influencing the penetration of antibacterial drugs into Gram-negative bacteria is explored in Chapter 7.

Lipoprotein is another major component of the outer membrane proteins. About one-third is linked to peptidoglycan and the remaining two-thirds are unattached, but form part of the membrane. The nature of the attachment of lipoprotein to the side chains of peptidoglycan is discussed later. About 1 in 12 of the peptide side chains is substituted in this way. This arrangement anchors the outer membrane to the peptidoglycan layer. The fatty acid chains of the lipoprotein presumably align themselves in the hydrophobic inner layer of the outer membrane and the protein moiety may possibly associate with matrix protein, reinforcing the pore structure.

Many other proteins with specialized functions have been identified in the outer membrane. Some of these are transport proteins, allowing access to molecules such as vitamin B_{12} or nucleosides which are too large to penetrate the pores of the membrane. Outer membrane proteins that contribute to the function of multidrug efflux pumps are described in Chapter 7.

2.3 Structure and biosynthesis of peptidoglycan

The structure and biosynthesis of peptidoglycan have special significance relative to the action of a number of important antibacterials and have been studied extensively. The biosynthesis of peptidoglycan was first worked out with *Staphylococcus aureus*. Although bacteria show many variations in peptidoglycan structure, the biosynthetic sequence in *Staphylococcus aureus* serves to illustrate the general features of the process. The biosynthesis may be conveniently divided into four stages.

2.3.1 Stage 1. Synthesis of UDP-*N*-acetylmuramic acid

The biosynthesis starts in the cytoplasm with two products from the normal metabolic pool, *N*-acetylglucosamine 1-phosphate and UTP (Figure 2.4). UDP-*N*-acetylglucosamine (I) is formed with the elimination of pyrophosphate. This nucleotide reacts with phosphoenolpyruvate by means of UDP-*N*-acetylglucosamine enolpyruvyl transferase to give the corresponding 3-enolpyruvyl ether (II). The pyruvyl group is then converted to lactyl by a reductase requiring NADPH, the product being UDP-*N*-acetylmuramic acid (III, UDPMurNAc). Muramic acid (3-*O*-D-lactyl-D-glucosamine) is a distinctive amino sugar derivative found only in the peptidoglycan of cell walls.

2.3.2 Stage 2. Building the pentapeptide side chain

Five amino acid residues are next added to the carboxyl group of the muramic acid nucleotide (Figure 2.5). Each step requires ATP and a specific enzyme. L-Alanine is added first. The next two residues are D-glutamic acid and then L-lysine. The lysine, however, is attached through its α-amino group to the γ-carboxyl group of the glutamic acid. The α-carboxyl group of the glutamic acid is amidated at a later stage in the biosynthesis, so the second amino acid residue is sometimes referred to as D-isoglutamine. The biosynthesis of the pentapeptide is completed by addition not of an amino acid but of a dipeptide, D-alanyl-D-alanine, which is separately synthesized. A racemase acting on L-alanine gives D-alanine, and a ligase then joins two molecules, giving the dipeptide. The completed UDP-*N*-acetylmuramyl intermediate (V) with its pendant peptide group will be referred to as the 'nucleotide pentapeptide'.

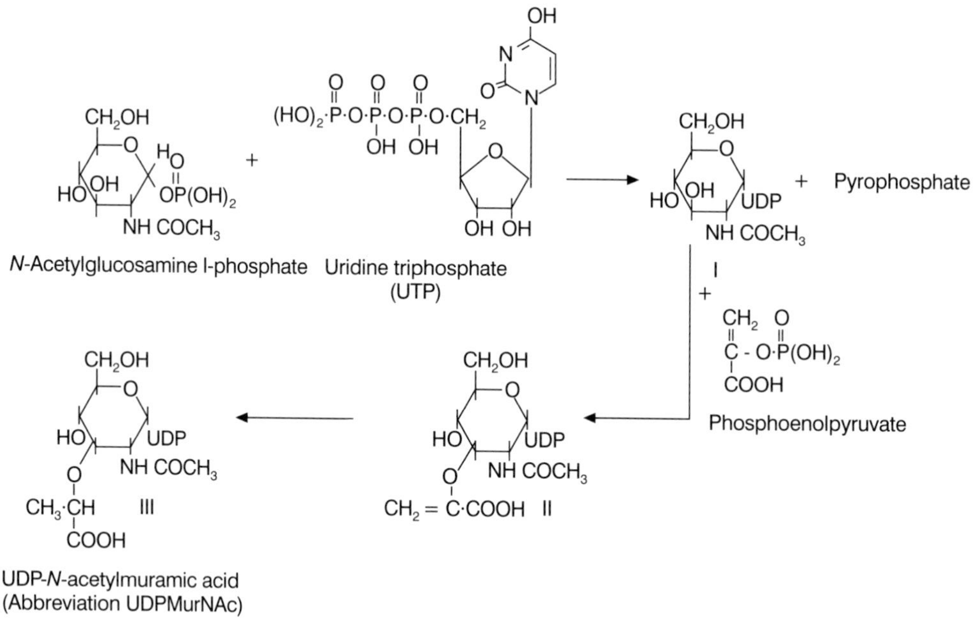

FIGURE 2.4 Peptidoglycan synthesis in *Staphylococcus aureus*. Stage 1: formation of UDP-*N*-acetylmuramic acid.

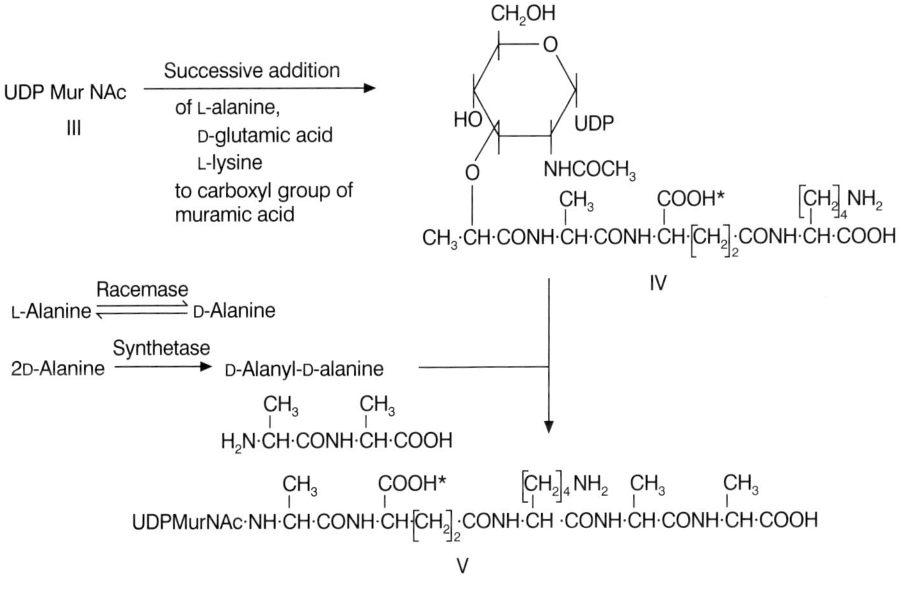

UDP Mur NAc III → Successive addition of L-alanine, D-glutamic acid, L-lysine to carboxyl group of muramic acid → IV

L-Alanine ⇌ (Racemase) D-Alanine

2D-Alanine → (Synthetase) D-Alanyl-D-alanine

$H_2N \cdot CH(CH_3) \cdot CONH \cdot CH(CH_3) \cdot COOH$

IV: (UDP-N-acetylmuramic acid, with CH_2OH, HO, UDP, $NHCOCH_3$, O-linked) $CH_3 \cdot CH \cdot CONH \cdot CH(CH_3) \cdot CONH \cdot CH(COOH^*) \cdot [CH_2]_2 \cdot CONH \cdot CH([CH_2]_4NH_2) \cdot COOH$

V: $UDPMurNAc \cdot NH \cdot CH(CH_3) \cdot CONH \cdot CH(COOH^*)[CH_2]_2 \cdot CONH \cdot CH([CH_2]_4NH_2) \cdot CONH \cdot CH(CH_3) \cdot CONH \cdot CH(CH_3) \cdot COOH$

UDP-*N*-acetylmuramyl pentapeptide

FIGURE 2.5 Peptidoglycan synthesis. Stage 2: formation of UDP-*N*-acetylmuramyl pentapeptide. Addition of each amino acid and the final dipeptide requires ATP and a specific enzyme. L-Lysine is added to the γ-carboxyl group of D-glutamic acid; the α-carboxyl group (marked *) is amidated at a later stage in the biosynthesis.

2.3.3 Stage 3. Membrane-bound reactions leading to a linear peptidoglycan polymer

The biosynthesis up to this point is cytoplasmic while the succeeding steps occur on membrane structures. The first membrane-associated step involves the formation of a pyrophosphate link between the nucleotide pentapeptide and undecaprenyl phosphate (the phosphate ester of a C_{55} isoprenoid alcohol) which is a component of the cytoplasmic membrane. In this reaction UMP is released and becomes available for reconversion to UTP which is needed in the first step of peptidoglycan biosynthesis (Figure 2.6). All subsequent reactions occurring while the intermediates are linked to undecaprenyl phosphate take place without release from the membrane. An essential step in this membrane-bound reaction sequence is the addition of a second hexosamine residue through a typical glycosidation by UDP-*N*-acetylglucosamine catalysed by a transglycosylase (Figure 2.6). The disaccharide (VII) is formed by a 1,4-β-linkage with liberation of UDP. The involvement of undecaprenyl phosphate is not unique to peptidoglycan biosynthesis. It is also concerned in the biosynthesis of the polysaccharide chain in the O-antigen produced by *Salmonella typhimurium* and in the formation of the polysaccharide elements of the lipopolysaccharides of Gram-negative bacteria; in Gram-positive bacteria it fulfils a similar role in the biosynthesis of teichoic acid or polysaccharides of the wall.

At about this point in the biosynthesis of *Staphylococcus aureus* peptidoglycan an extending group is added to the ϵ-amino group of the lysine unit in the nucleotide pentapeptide. Glycine and a glycine-specific transfer (t)RNA are involved in this process during which a pentaglycine group is added. The reaction, unlike the tRNA reactions in protein biosynthesis, does not require the presence of ribosomes; the five glycine units are added successively to the lysine from the nitrogen end (the

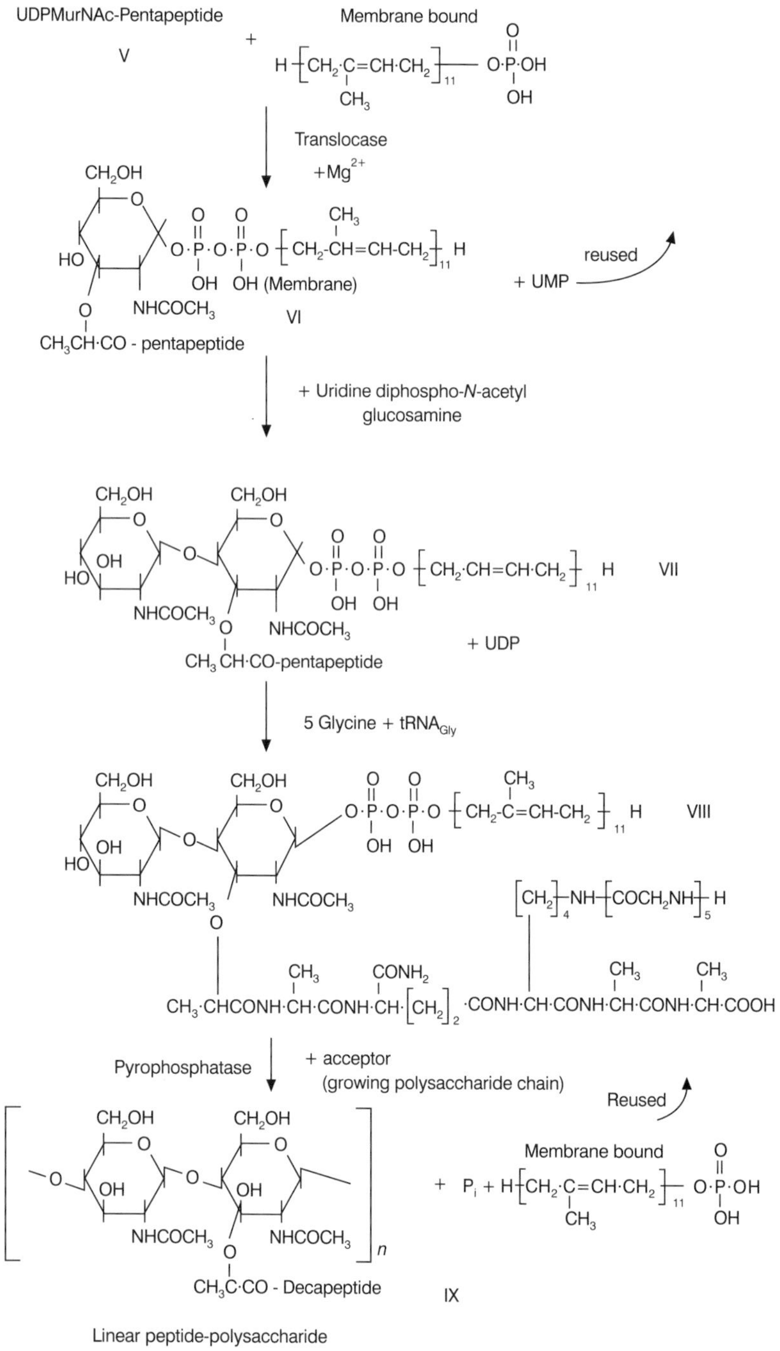

UDPMurNAc-Pentapeptide V + Membrane bound $H[CH_2 \cdot C(CH_3){=}CH \cdot CH_2]_{11}$—O·P(O)(OH)·OH

Translocase $+Mg^{2+}$

CH_2OH … HO … O·P(O)(OH)·O·P(O)(OH)·O $[CH_2 \cdot C(CH_3){=}CH \cdot CH_2]_{11}$ H (Membrane)

$NHCOCH_3$

$CH_3CH \cdot CO$ - pentapeptide

VI

+ UMP — reused

+ Uridine diphospho-*N*-acetyl glucosamine

CH_2OH CH_2OH … OH … HO … O·P(O)(OH)·O·P(O)(OH)·O $[CH_2 \cdot CH{=}CH \cdot CH_2]_{11}$ H VII

$NHCOCH_3$ $NHCOCH_3$

$CH_3CH \cdot CO$-pentapeptide

+ UDP

5 Glycine + $tRNA_{Gly}$

CH_2OH CH_2OH … OH … HO … O·P(O)(OH)·O·P(O)(OH)·O $[CH_2 \cdot C(CH_3){=}CH \cdot CH_2]_{11}$ H VIII

$NHCOCH_3$ $NHCOCH_3$

$[CH_2]_4$—NH—$[COCH_2NH]_5$—H

$CH_3 \cdot CHCONH \cdot CH(CH_3) \cdot CONH \cdot CH(CONH_2) \cdot [CH_2]_2 \cdot CONH \cdot CH \cdot CONH \cdot CH(CH_3) \cdot CONH \cdot CH(CH_3) \cdot COOH$

Pyrophosphatase + acceptor (growing polysaccharide chain)

$[$—O— CH_2OH … OH … CH_2OH … OH … $NHCOCH_3$ $NHCOCH_3$ $]_n$

$CH_3C \cdot CO$ - Decapeptide

Linear peptide-polysaccharide

IX

+ P_i + Membrane bound $H[CH_2 \cdot C(CH_3){=}CH \cdot CH_2]_{11}$ O·P(O)(OH)·OH — Reused

FIGURE 2.6 Peptidoglycan synthesis. Stage 3: formation of the linear peptidoglycan. The structure of decapeptide side chain is shown in VIII.

reverse direction to protein biosynthesis). Since the resultant product (VIII) has 10 amino acid units, it is referred to as the disaccharide decapeptide. This still has a free terminal amino group. In the biosynthesis of peptidoglycans in certain other bacterial species, for example in *Escherichia coli*, no extending group is added. Later reactions then involve the ϵ-amino group of lysine (or equivalent diamino acid) instead of the terminal amino group of glycine. Also, during the membrane-bound stage in the biosynthesis of *Staphylococcus aureus* peptidoglycan the carboxyl group of D-glutamic acid is amidated by a reaction with ammonia and ATP.

The final reaction in this stage is the attachment of the disaccharide decapeptide (VIII) to an 'acceptor', usually regarded as the growing linear polymer chain. In this reaction the disaccharide with its decapeptide side chain forms a β-linkage from the 1-position of the *N*-acetylmuramic acid residue to the 4-hydroxyl group of the terminal *N*-acetylglucosamine residue in the growing polysaccharide chain. This reaction occurs outside the cytoplasmic membrane and the disaccharide–decapeptide linked to the undecaprenyl phosphate moves across the membrane to gain access to the acceptor on the external face of the membrane. The released undecaprenyl pyrophosphate is reconverted by a specific pyrophosphatase to the corresponding phosphate, ready for another cycle of the membrane-bound part of the synthesis. The growth of the glycan chains thus occurs by successive addition of disaccharide units.

2.3.4 Stage 4. Cross-linking

The linear peptidoglycan (IX) formed in stage 3 contains many polar groups which make it soluble in water. It lacks mechanical strength and toughness. These attributes are introduced in the final stage of biosynthesis by cross-linking, a process well known in the plastics industry for producing similar results in synthetic linear polymers. The mechanism involved in cross-linking is a transpeptidation reaction requiring no external supply of ATP or similar compounds. In *Staphylococcus aureus* the transpeptidation occurs between the terminal amino group of the pentaglycine side chain and the peptide amino group of the terminal D-alanine residue of another peptide side chain; D-alanine is eliminated and a peptide bond is formed (Figure 2.7). In *Staphylococcus aureus* peptidoglycan the cross-linking is quite extensive and up to 10 peptide side chains may be bound together by bridging groups. Since the linear polymers themselves are very large, it is likely that the whole of the peptidoglycan in a Gram-positive bacterium is made up of units covalently bound together. This gigantic bag-shaped molecule has been called a 'sacculus'. There is also a mechanism for constantly breaking it down and re-forming it to allow for cell growth and division. Peptidoglycan hydrolases which hydrolyse the polysaccharide chains of peptidoglycan and others attacking the peptide cross-links exert this essential catabolic activity during cell growth.

2.3.5 Variations in peptidoglycan structure

Many variations are found in peptidoglycan structure between one species of bacteria and another or even between strains of the same species, and only a general account is possible here. All peptidoglycans have the same glycan chain as in *Staphylococcus aureus* except that the glucosamine residues are sometimes *N*-acylated with a group other than acetyl. The peptide side chains always have four amino acid units alternating L-, D-, L-, D- in configuration. The second residue is always D-glutamic acid, linked through its γ-carboxyl group, and the fourth invariably D-alanine. The peptidoglycan from *Staphylococcus aureus* (Type A2) is characteristic of many Gram-positive cocci. Peptidoglycans of this group, and the related Types A3 and A4, have similar tetrapeptide side chains but vary in their bridging groups. The amino acids in the bridge are usually glycine, alanine, serine or

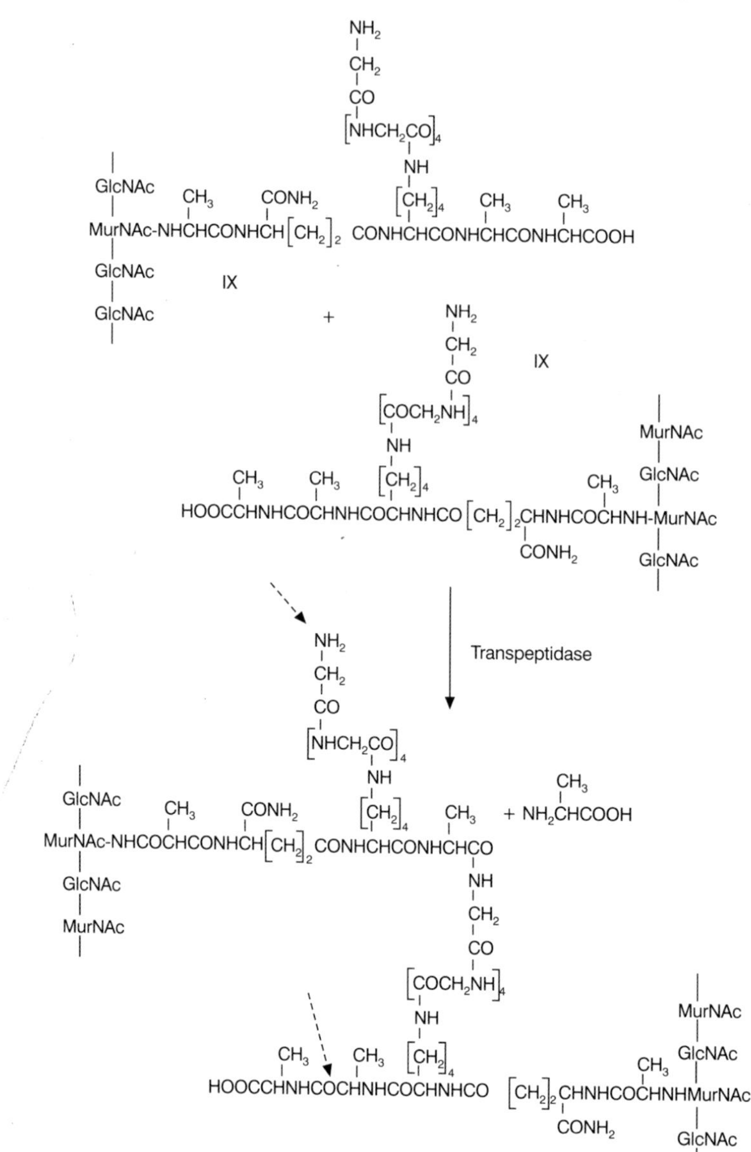

FIGURE 2.7 Peptidoglycan synthesis. Stage 4: cross-linking of two linear peptidoglycan chains. The linear polymers have the structure IX (Figure 2.6) GlcNac: *N*-acetylglucosaminyl residue. Broken arrows show points at which further cross-links may be formed with other polymer chains. MurNAc: *N*-acetylmuramyl residue.

threonine and the number of residues can vary from one to five. In Type Al peptidoglycans the L-lysine of the Type II peptide side chain is usually replaced by *meso*-2,6-diaminopimelic acid, and there is no bridging group. Cross-linking occurs between the D-alanine of one side chain and the 6-amino group of the diaminopimelic acid of another. This peptidoglycan type is characteristic of many rod-shaped bacteria, both the large family of Gram-negative rods and the Gram-positive bacilli.

In the less common Type B peptidoglycans cross-linkage occurs between the α-carboxyl group of the D-glutamic acid of one peptide side chain and the D-alanine of another, through a bridge containing a basic amino acid.

2.3.6 Cross-linking in Gram-negative bacteria

In contrast to the multiple random cross-linkage of peptidoglycan which is found in the Gram-positive cocci, the peptidoglycan of *Escherichia coli* and similar Gram-negative rods has on average only a single cross-link between one peptide side chain and another. These bacteria contain, besides the transpeptidases concerned in cross-linkage, other enzymes known as DD-carboxypeptidases which specifically remove D-alanine from a pentapeptide side chain. Carboxypeptidase I is specific for the terminal D-alanine of the pentapeptide side chain, whereas carboxypeptidase II acts on the D-alanine at position 4 after the terminal D-alanine has been removed. DD-Carboxypeptidase I therefore limits the extent of cross-linking.

The peptidoglycan sacculus determines the overall shape of the cell and the peptidoglycan is laid down with a definite orientation in which the polysaccharide chains run perpendicular to the main axis of rod-shaped organisms (e.g. *Escherichia coli*).

2.3.7 Penicillin-binding proteins (PBPs)

The enzymes involved in linking the disaccharide deca- or pentapeptide to the growing linear peptidoglycan and the subsequent cross-linking reaction are referred to as PBPs (see Table 2.1) because of the ability of penicillin to bind to them covalently. As we shall see when the PBPs are more fully described later in the chapter, this latter reaction, which inactivates the cross-linking function although not the transglycosylase activity, is central to the antibacterial activity of all β-lactam antibiotics.

2.3.8 Attachments to peptidoglycans

Within the cell wall the polymeric peptidoglycan is usually only part of a larger polymer. In Gram-positive cocci it is linked to an acidic polymer, often a teichoic acid (Figure 2.8). The point of attachment is through the 6-hydroxyl group of muramic acid in the glycan chain. Only a small fraction of the muramic acid residues are thus substituted. In *Staphylococcus aureus* cell walls teichoic acid is joined to peptidoglycan by a linking unit comprising three glycerol 1-phosphate units attached to the 4-position of *N*-acetylglucosamine which engages through a phosphodiester group at position 1 with the 6-hydroxyl group of muramic acid. This type of linkage seems to occur with polymers other than teichoic acid, e.g. with poly(*N*-acetylglucosamine 1-phosphate) in a *Micrococcus* species. The acid-labile *N*-acetylglucosamine l-phosphate linkage and the alkali-labile phosphodiester linkage at position 4 explain the ease with which teichoic acid can be split off from peptidoglycan. Within the cell wall the synthesis of teichoic acid is closely associated with that of peptidoglycan.

In the Gram-positive mycobacteria the peptidoglycan carries quite a different polymeric attachment. Arabinogalactan is attached to the 6-position of some of the *N*-glycolylmuramic acid residues of the glycan chain through a phosphate ester group. Mycolic acids (complex, very long-chain fatty acids) are attached by ester links to the C-5 position of arabinose residues of the arabinogalactan. The mycobacterial cell wall thus has a high lipid content.

In *Escherichia coli* and related bacteria the peptidoglycan carries a lipoprotein as a substituent (Figure 2.9). The lipoprotein consists of a polypeptide chain of 58 amino acid units of known sequence with lysine at the C-terminus and cysteine at the N-terminus. This is attached to the 2-carboxyl group of *meso*-2,6-diaminopimelic acid in a peptide side chain of *Esherichia coli* peptidoglycan which has lost both D-alanine groups. Attachment is by an amide link with the ε-amino group in the terminal lysine of the polypeptide. At the opposite end of the polypeptide chain the cysteine amino group carries a long-chain fatty acid joined as an amide, and its sulphur atom forms a thioether link with a long-chain diacylglycerol.

Lipoprotein occurs in enteric bacteria other than *Escherichia coli*, but it may not be common to

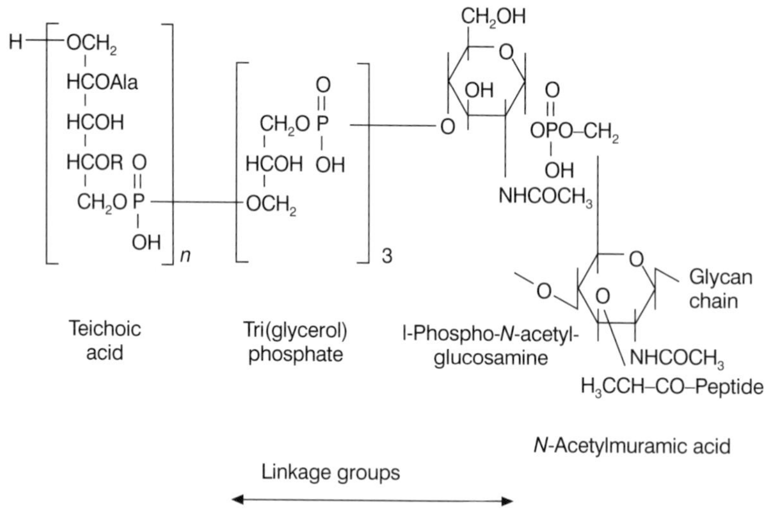

FIGURE 2.8 Teichoic acid and its linkage to peptidoglycan in the wall of *Staphylococcus aureus*.

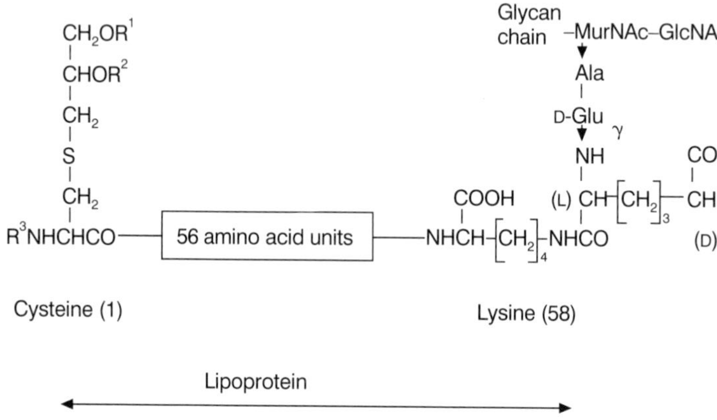

FIGURE 2.9 Lipoprotein and its linkage to peptidoglycan in the envelope of *Escherichia coli*.

all Gram-negative bacteria, although small amounts have been detected in *Proteus mirabilis*.

2.4 Antibiotics that inhibit peptidoglycan biosynthesis

The conclusion that a particular antibiotic owes its antibacterial activity to interference with peptidoglycan biosynthesis rests on several lines of evidence:

1. Bacteria suspended in a medium of high osmotic pressure are protected from concentrations of the antibiotic that would cause lysis and death in a normal medium. Under these conditions the cells lose the shape-determining action of the peptidoglycan and become spherical; they are then known as spheroplasts. These retain an undamaged cytoplasmic membrane but their wall is deficient or considerably modified. Spheroplasts are in principle viable and if the antibiotic is removed they can divide and produce progeny with normal walls.
2. Several species of bacteria have walls containing no peptidoglycan. These include the mycoplasmas, the halophilic bacteria tolerant of high salt concentrations and bacteria in the L-phase, where the normal wall structure is greatly mod-

ified. If a compound inhibits the growth of common bacteria but fails to affect bacteria of these special types, it probably owes its activity to interference with peptidoglycan synthesis.

3. Subinhibitory concentrations of these antibiotics often cause accumulation in the culture medium of uridine nucleotides of *N*-acetylmuramic acid with varying numbers of amino acid residues attached, which represent intermediates in the early stages of peptidoglycan biosynthesis. When an antibiotic causes a block at an early point in the reaction sequence it is not surprising to find accumulation of the intermediates immediately preceding the block. However, quantities of muramic acid nucleotides are also found in bacteria treated with antibiotics known to affect later stages in peptidoglycan biosynthesis. It seems that all the biosynthetic steps associated with the membrane are closely interlocked and inhibition of any one of them leads to accumulation of the last water-soluble precursor, UDP-*N*-acetylmuramyl pentapeptide (V, Figure 2.5).

2.4.1 Bacitracin

Bacitracin, a polypeptide antibiotic (Figure 2.10), is too toxic for systemic administration but is sometimes used topically to kill Gram-positive bacteria. Its effect on peptidoglycan biosynthesis depends upon its ability to bind specifically to polyprenyl pyrophosphates in the presence of magnesium ions. In the formation of the linear peptidoglycan (IX, Figure 2.7) the membrane-bound undecaprenyl pyrophosphate is released. Normally this is converted by a pyrophosphatase to the corresponding phosphate which thus becomes available for reaction with another molecule of UDPMurNAc-pentapeptide (V). Formation of the complex between the lipid pyrophosphate with bacitracin blocks this process and so eventually halts the synthesis of peptidoglycan. Bacitracin also inhibits sterol biosynthesis in animal tissues through complex formation with intermediates such as farnesyl pyrophosphate.

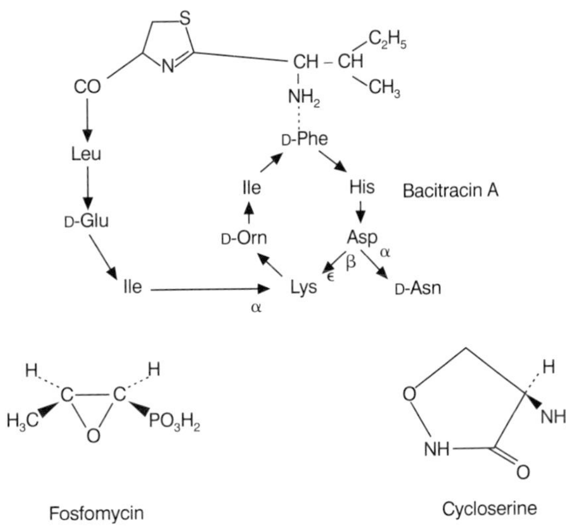

FIGURE 2.10 Antibiotics affecting intracellular stages in the biosynthesis of peptidoglycan.

2.4.2 Fosfomycin (phosphonomycin)

This antibiotic has the very simple structure shown in Figure 2.10. It acts on infections caused by both Gram-positive and Gram-negative bacteria but although its toxicity is low it has achieved only limited use in clinical practice. Its inhibitory action is exerted on the first step of peptidoglycan biosynthesis, namely the condensation of UDP-*N*-acetylglucosamine (I) with phosphoenolpyruvate (PEP) by means of UDP-*N*-acetylglucosamine enolpyruvyl transferase (otherwise known as MurA) giving the intermediate (II) that subsequently yields UDP-*N*-acetylmuramic acid (III) on reduction (Figure 2.4). Fosfomycin inactivates MurA by reacting covalently with an essential cysteine residue (Cys-115) at the active centre of the enzyme to form the thioester illustrated in Figure 2.11. This reaction is time-dependent and is facilitated by UDP-*N*-acetylglucosamine which appears to 'chase' the other substrate (PEP) from the active site and to promote a conformational change in the enzyme. Both these effects are believed to expose the nucleophilic Cys-115 for reaction with the epoxide moiety of fosfomycin. Recently the three-dimensional structure of MurA (from

Escherichia coli) complexed with UDP-*N*-acetylglucosamine and fosfomycin was determined by X-ray crystallography. The analysis confirmed the covalent interaction of the antibiotic with Cys-115 and also revealed that there are hydrogen bonds between the antibiotic and both the enzyme and UDP-*N*-acetylglucosamine.

2.4.3 Cycloserine

This antibiotic also has a simple structure (Figure 2.10). It is active against a number of bacterial species but has found little clinical use because of the central nervous system disturbances that are sometimes experienced by patients. Cycloserine shows the usual effects that characterize compounds acting on peptidoglycan biosynthesis. Thus when cultures of *Staphylococcus aureus* are grown with subinhibitory concentrations of cycloserine, the peptidoglycan precursor (IV) (Figure 2.5) accumulates in the medium, suggesting a blockage in the biosynthesis immediately beyond this point.

In fact cycloserine inhibits both alanine racemase and D-alanyl-D-alanine ligase, the two enzymes concerned in making the dipeptide for completion of the pentapeptide side chain. Molecular models reveal that cycloserine is structurally related to one possible conformation of D-alanine, so its inhibitory action on these enzymes appears to be a classical example of isosteric interference. The observation that the action of cycloserine is specifically antagonized by the addition of D-alanine to the growth medium also supported the postulated site of action. The affinity of cycloserine for the ligase is much greater than that of the natural substrate, the ratio of K_m to K_i being about 100. In a compound acting purely as a competitive enzyme inhibitor, this sort of K_m/K_i ratio is probably essential for useful antibacterial activity. The greater affinity of cycloserine for the enzyme may be connected with its rigid structure. This could permit a particularly accurate fit to the active centre of the enzyme, either in the state existing when the enzyme is uncombined with its substrate or in a modified conformation which is assumed during the normal enzymic reaction. The three-dimensional structure of the D-Ala–D-Ala ligase is now available and it would be interesting, using computer molecular modelling techniques, to see whether the cycloserine structure can be 'docked' into the enzyme structure according to this concept of inhibition. Rigid structures of narrow molecular specificity are common among antimicrobial agents and similar considerations may apply to other types of action; this theme will recur in later sections.

Cycloserine enters the bacterial cell by active transport (see Chapter 7). This allows the antibiotic to reach higher concentrations in the cell than in the medium and adds considerably to its antibacterial efficacy.

FIGURE 2.11 Fosfomycin inactivates UDP-*N*-acetylglucosamine enolpyruvoyl transferase (MurA) by reacting with the essential cysteine residue (Cys-115) at the active centre of the enzyme to form a thioester.

2.4.4 Glycopeptide antibiotics: vancomycin and teichoplanin

Vancomycin (Figure 2.12) has been known for many years but its clinical importance has emerged only relatively recently with the spread of methicillin-resistant staphylococci. The use of vancomycin and the structurally related teichoplanin is steadily increasing because of their value in this therapeutic area. Vancomycin is also useful against intestinal infections due to *Clostridium difficile.* This organism sometimes multiplies and produces toxins when the usual gut flora have been largely eliminated by the use of broad-spectrum antibiotics. The action of vancomycin and teichoplanin depends on their ability to bind specifically to the terminal D-alanyl-D-alanine group on the peptide side chain of the membrane-bound intermediates in

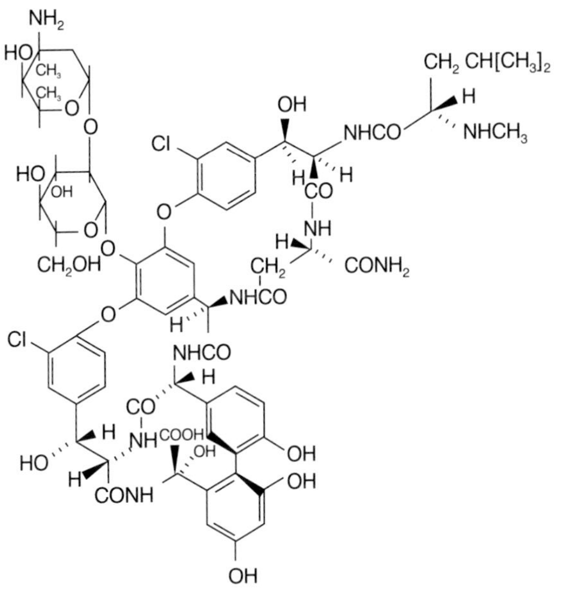

FIGURE 2.12 Vancomycin, an antibiotic that is increasingly important in the treatment of infections caused by drug-resistant staphylococci.

peptidoglycan synthesis (compounds VI–IX in Figure 2.6). It is important to note that this interaction occurs on the outer face of the cytoplasmic membrane and that these antibiotics probably do not enter the bacterial cytoplasm. The formation of the complex between vancomycin and D-alanyl-D-alanine blocks the transglycosylase involved in the incorporation of the disaccharide-peptide into the growing peptidoglycan chain and also the DD-transpeptidases and DD-carboxypeptidases for which the D-alanyl-D-alanine moiety is a substrate. Thus both peptidoglycan chain extension and cross-linking are inhibited by the glycopeptide antibiotics.

The nature of the interaction between vancomycin and the terminal D-alanyl-D-alanine of the peptide side chain is especially interesting. The side chains of the amino acids of the hexapeptide backbone of vancomycin are extensively cross-linked to form a relatively concave cleft into which the D-alanyl-D-alanine entity binds non-covalently. Furthermore, NMR and X-ray crystallographic studies show that vancomycin spontaneously forms a dimeric structure which enables the antibiotic to bind to two D-alanyl-D-alanine peptides attached either to the disaccharide-peptide precursor or to adjacent growing peptidoglycan strands (Figure 2.13). Although the dimerization of vancomycin is thought to facilitate the antibiotic activity of the molecule, it is interesting to note that teichoplanin does not form dimers and this may be the reason why teichoplanin is effective against some bacteria that have become resistant to vancomycin.

2.4.5 Penicillins, cephalosporins and other β-lactam antibiotics

Penicillin was the first antibiotic to be used in medicine. It is one of a group of compounds known as β-lactam antibiotics which are unrivalled in the treatment of bacterial infections. Their only serious defects include their ability to cause immunologic sensitization in a small proportion of patients, a side effect which prevents their use in those affected, and the frequency of emergence of bacteria resistant to β-lactams. The original penicillins isolated directly from mould fermentations were mixtures of compounds having different side chains. Addition of phenylacetic acid to the fermentation medium improved the yield of penicillin and ensured that the product was substantially a single compound known as penicillin G or benzylpenicillin (Figure 2.14). The first successful variant was obtained by replacing phenylacetic acid by phenoxyacetic acid as the added precursor. This gave phenoxymethylpenicillin or penicillin V (Figure 2.14). The main advantage of this change was an improvement in the stability of the penicillin towards acid. The ready inactivation of penicillin G at low pH limited its usefulness when it was given by mouth since a variable and often considerable fraction of the antibacterial activity was lost in the stomach. Penicillin V thus improved the reliability of oral dosing. These early penicillins, produced directly by fermentation, were intensely active against Gram-positive infections and gave excellent

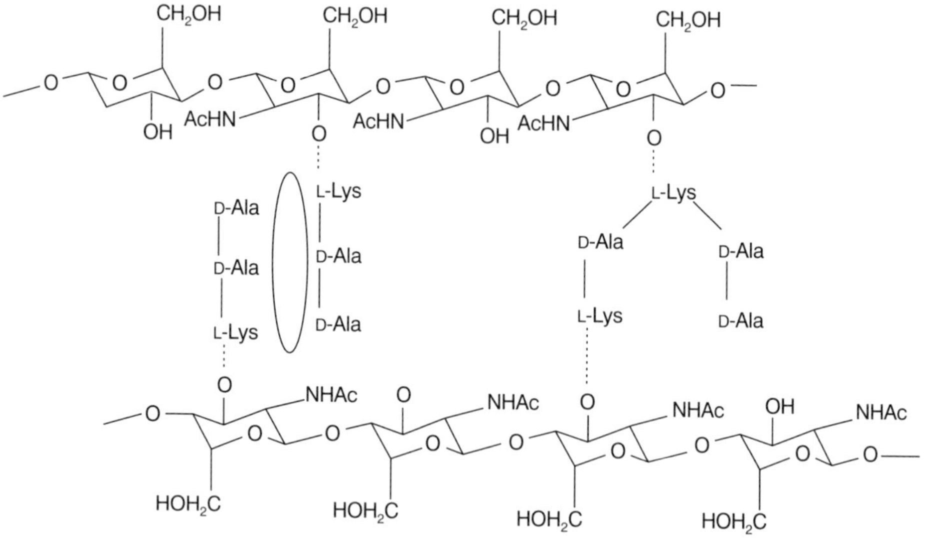

FIGURE 2.13 A likely mode of the simultaneous interaction of dimerized vancomycin molecules (represented by the open ellipse) with the terminal D-alanyl-D-alanine residues of adjacent growing peptidoglycan chains. The interaction is believed to underlie the inhibition of the transglycolase involved in peptidoglycan biosynthesis. Only the last three amino acids of the pentapeptide side chains are shown. (Adapted with permission from M. Schäfer *et al.* (1996) *Structure,* **4**, 1509.)

results in streptococcal and staphylococcal infections and in pneumonia. They were also very active against Gram-negative infections caused by gonococci and meningococci but were much less active against the more typical Gram-negative bacilli.

A further advance in the versatility of the penicillins was achieved by workers of the Beecham (now SmithKlineBeecham) group with the development of a method for the chemical modification of the penicillin molecule. Bacterial enzymes were found that remove the benzyl side chain from penicillin G, leaving 6-aminopenicillanic acid which could be isolated and then acylated by chemical means. This discovery opened the way to the production of an almost unlimited number of penicillin derivatives, some of which have shown important changes of properties compared with the parent penicillin. Three types of improvement have been achieved. The value of increased stability has already been mentioned, and some semi-synthetic penicillins show this property. Some modified penicillins (e.g. methicillin and cloxacillin, Figure 2.14) are much less susceptible to attack by β-lactamase, an enzyme which converts penicillin to the antibacterially inactive penicilloic acid and which gives rise to the most common form of resistance to penicillin (Chapter 9).

The discovery of the β-lactamase inhibitor, clavulanic acid (see Figure 2.15) which is a β-lactam itself but without useful antibiotic activity, provided an opportunity to co-administer this agent with β-lactamase-sensitive compounds such as amoxycillin (Figure 2.14) in mixtures such as augmentin (a 1:1 mixture of amoxycillin and clavulanic acid) and timentin (a 1:1 mixture of ticarcillin and clavulanic acid).

Another striking change brought about by chemical modification of the penicillin side chain was an increase of activity against Gram-negative bacteria, a property found in ampicillin and amoxycillin. The increase in this type of activity is accompanied by a lessening of activity towards Gram-positive bacteria. Ampicillin is one of the most widely used antibacterial agents. In mecilli-

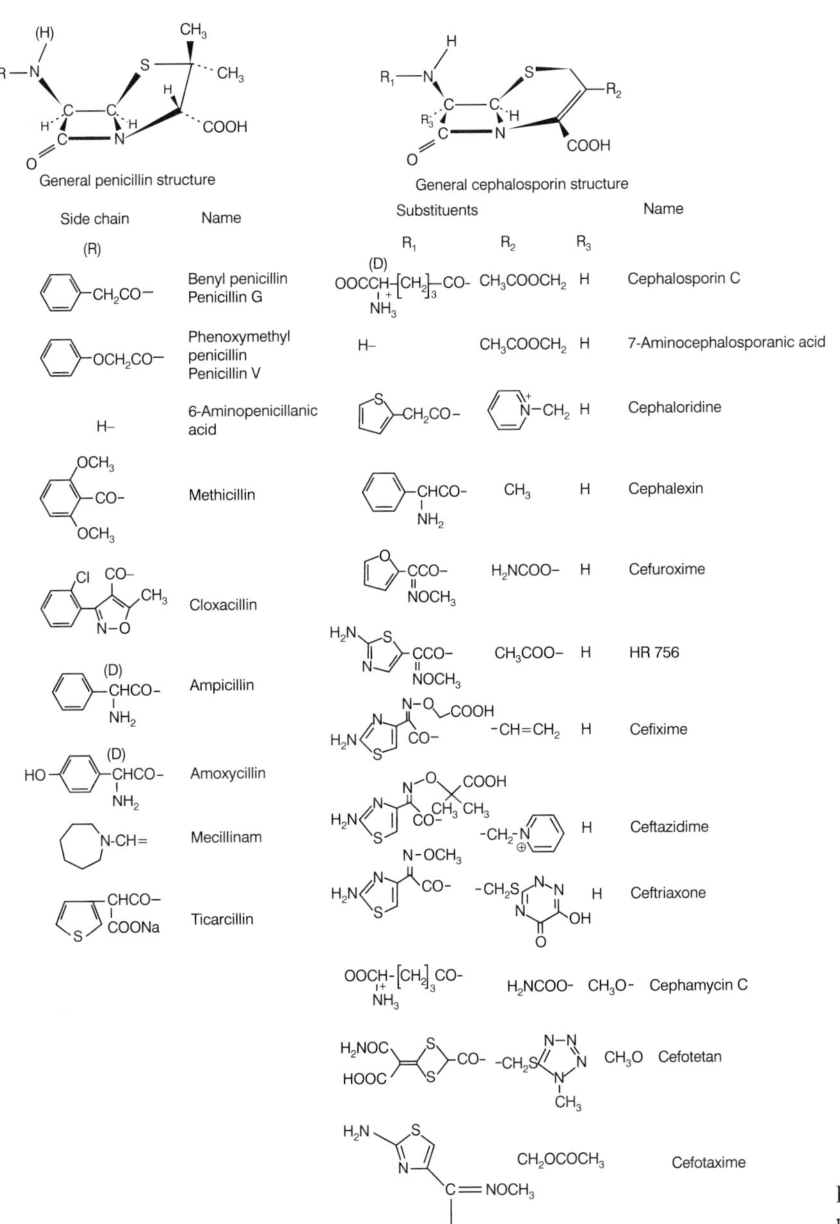

FIGURE 2.14 Representative penicillins and cephalosporins.

nam (Figure 2.14) the side chain is attached by an azomethine link rather than the usual amide bond. Here the activity spectrum of the original penicillin molecule has been completely reversed. This compound is highly active against Gram-negative bacteria but requires 50 times the concentration for an equal effect on Gram-positive organisms. It can be used in the treatment of typhoid fever, which is caused by the Gram-negative bacterium *Salmonella typhi*.

Cephalosporin C (Figure 2.14), isolated from a different organism from that used for penicillin production, was shown to have a similar structure in its nucleus to the penicillins. The biogenesis of the nuclei in these two classes of antibiotics is now known to be identical except that in cephalosporin biosynthesis the thiazolidine ring of the penicillin nucleus undergoes a specific ring expansion to form the dihydrothiazine ring of the cephalosporin nucleus. Besides this similarity in structure and biogenesis, cephalosporin C and its derivatives act on peptidoglycan cross-linking in the same way as the penicillins. Cephalosporin C itself is not a useful antibacterial drug but, like the penicillins, it is amenable to chemical modification. Enzymic removal of the side chain gives 7-aminocephalosporanic acid which can be chemically acylated to give new derivatives. A second change in the molecule can also be made by a chemical modification of the acetoxy group of cephalosporin C. The first successful semi-synthetic cephalosporin was cephaloridine. Many others have followed, a selection of some of the best known being shown in Figure 2.14. Most are only effective when given by injection, but cephalexin and cefixime can be given by mouth. Cefuroxime is unaffected by many of the common β-lactamases and can be used against bacterial strains which are resistant to other β-lactam antibiotics; it can be useful in infections due to *Neisseria* or *Haemophilus*. The related compound cefotaxime has enjoyed considerable success. Other agents such as ceftazidime and ceftriaxone are finding favour because of the former's improved antipseudomonal activity and the latter's enhanced half-life in the body, which permits a more convenient dosing schedule, for example, once or twice daily.

The cephamycins resemble the cephalosporins, but have a methoxy group in place of hydrogen at position 7. Cefotetan (Figure 2.14) is a semi-synthetic derivative of cephamycin C. The cephamycin derivatives are not readily attacked by β-lactamases and have advantages over the cephalosporin derivatives with activity against *Proteus* and *Serratia* species.

The enormous success of the penicillins and cephalosporins has stimulated a search for other naturally occurring β-lactam compounds. These have been found in a variety of micro-organisms. Some of the most interesting are shown in Figure 2.15. In the carbapenem, thienamycin, the sulphur atom is not part of the ring, but is found in the side chain. This compound is remarkable for its high potency, broad antibacterial spectrum and resistance to β-lactamase attack, but it is both chemically unstable and susceptible to degradation by a dehydropeptidase found in the kidneys. The *N*-formimidoyl derivative of thienamycin imipenem is chemically more stable but must be administered as a 1:1 mixture with cilastatin (Figure 2.15), an inhibitor of the renal peptidase. A further development in the carbapenem series has been the appearance of meropenem (Figure 2.15), which is not readily degraded by the renal peptidase and can therefore be administered a single agent.

Other β-lactams include the monobactams (e.g. sulfazecin, Figure 2.15) named as *mono*cyclic *bac*terial β-lac*tams*, which are derived from bacteria and represent the simplest β-lactam structures so far discovered with antibacterial activity. Many semi-synthetic derivatives have been made and exhibit excellent anti-Gram-negative activity with much weaker activity against Gram-positive bacteria. In contrast, the monocyclic nocardicins (Figure 2.15) appear to offer less activity and are more of historic, rather than clinical, interest.

Mode of action of penicillins and cephalosporins

As with many other antibiotics, early attempts to discover the biochemical action of penicillin led to conflicting hypotheses. Gradually it became accepted that the primary site of action lay in the production of cell wall material, and more specifically in the biosynthesis of peptidoglycan.

Evidence for this site of action rests on several different types of experiment. *Staphylococcus aureus* cells were pulse-labelled with [^{14}C]glycine, and peptidoglycan was isolated from their walls after growth for a further period in unlabelled medium. The labelled glycine entered the pentaglycyl 'extending group'. The polysaccharide backbone

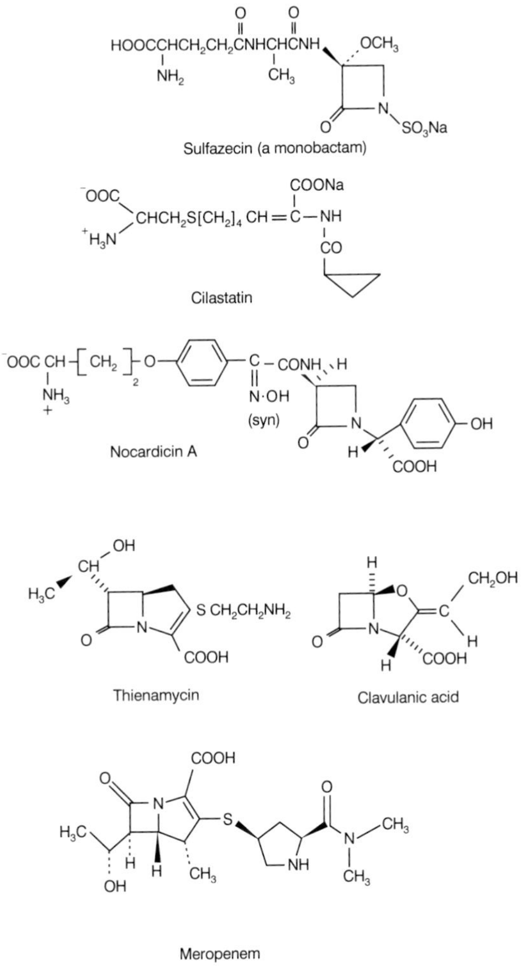

Sulfazecin (a monobactam)

Cilastatin

Nocardicin A

Thienamycin

Clavulanic acid

Meropenem

FIGURE 2.15 Additional β-lactam compounds and also cilastatin, an inhibitor of mammalian metabolism of thienamycin. Clavulanic acid is an inhibitor of serine-active-site β-lactamases.

of the peptidoglycan was then broken down by an *N*-acetylmuramidase, leaving the individual muramyl peptide units linked together only by their pentaglycine peptide chains. After the products were separated by gel chromatography, radioactivity was found in a series of peaks of increasing molecular weight, representing the distribution of the pulse of [^{14}C]glycine among peptide-linked oligomers of varying size. A parallel experiment performed in the presence of penicillin showed the radioactivity to be associated largely with a single peak of low molecular weight, presumably the uncross-linked muramyl peptide unit, with much less radiolabel in the oligomers. The penicillin had thus inhibited the peptide cross-linking.

In another experiment, 'nucleotide pentapeptide' was prepared with [^{14}C]alanine. This was used as a substrate for an enzyme preparation from *Escherichia coli* in the presence of UDP-*N*-acetylglucosamine. This system carried out the entire biosynthesis of peptidoglycan, including the final stage of cross-linking. Peptidoglycan was obtained as an insoluble product containing ^{14}C from the penultimate D-alanine of the substrate; the terminal D-[^{14}C]alanine was released into the medium, partly from the transpeptidase cross-linking reaction and partly from a carboxypeptidase that removed terminal D-alanine residues from cross-linked products. In a parallel experiment penicillin was added at a concentration that would inhibit growth of *Escherichia coli*. Biosynthesis of peptidoglycan then proceeded only to the stage of the linear polymer (IX, Figure 2.6) which was isolated as a water-soluble product of high molecular weight labelled with ^{14}C. No D-[^{14}C]alanine was liberated because the penicillin suppressed both the cross-linking transpeptidase reaction and the action of DD-carboxypeptidase.

The understanding of the mechanism of β-lactam action has been considerably advanced by the observation that there are groups of proteins to which β-lactams become covalently bound – the penicillin-binding proteins (PBPs) mentioned earlier.

The PBPs (Table 2.1) fall into two major groups of high (≥60 kDa) and low (≤49 kDa) molecular mass, respectively. The high molecular mass PBPs consist of three modules, one of which has a serine residue that is attacked by penicillin to form a covalent penicilloyl–protein complex. The penicillin-binding module is fused to the carboxy end of a non-penicillin-binding component on the outer surface of the cytoplasmic membrane to which it is anchored by the third, amino-terminal transmembrane module. Some high molecular mass PBPs have both transglycosylase and

TABLE 2.1 Properties of penicillin-binding proteins of *Escherichia coli*

Protein no.	*Molecular wt (kDa)*	*Enzyme activities*	*Function*
1a	91	Transpeptidase Transglycosylase	Peptidoglycan cross-linking
1b	91	Transpeptidase Transglycosylase	Peptidoglycan cross-linking
2	66	Transpeptidase Transglycosylase	Peptidoglycan cross-linking
3	60	Transpeptidase Transglycosylase	Peptidoglycan cross-linking
4	49	DD-Carboxypeptidase	Limitation of peptidoglycan cross-linking
5	41	DD-Carboxypeptidase	Limitation of peptidoglycan cross-linking
6	40	DD-Carboxypeptidase	Limitation of peptidoglycan cross-linking

transpeptidase activities, borne respectively by the non-penicillin-binding and penicillin-binding modules. PBPs 1a and 1b of *Escherichia coli* exemplify this bifunctional type and are described as Class A PBPs. The Class B *Escherichia coli* PBPs 2 and 3 also have both enzymic functions, although in this case the transglycosylase activity is not found on the non-penicillin-binding module and its precise location remains to be determined.

In *Escherichia coli*, PBPs 1a and 1b provide the key enzyme activities involved in peptidoglycan synthesis, while PBPs 2 and 3 are apparently involved with the remodelling of the peptidoglycan sacculus, which is necessary during septation and cell division. These four proteins are the principal targets for the action of β-lactam antibiotics against *Escherichia coli* and many other bacteria. Different β-lactams exhibit different affinities for the various PBPs and these can, in turn, be correlated with different morphological effects (Figure 2.16). Drugs which bind most strongly to PBPs 1a and 1b cause cell lysis at the lowest antibacterial concentration. Other compounds such as the cephalosporin, cephalexin, bind more strongly to PBP3 and inhibit septation, leading to the formation of filaments, which are greatly elongated cells (Figure 2.16c).

This distinction in the action of β-lactams associated with differences in binding suggests that different mechanisms may be involved in synthesis of peptidoglycan for cell extension compared with its synthesis for septum formation. Another variation is found with mecillinam, which binds almost exclusively to PBP2, and causes cells to assume an abnormal ovoid shape (Figure 2.16b). Cells overproducing PBP2 have enhanced amounts of cross-linked peptidoglycan and are very sensitive to mecillinam.

Unlike the high molecular mass PBPs the low molecular mass PBPs, which include PBPs 4, 5 and 6 in *Escherichia coli*, are monofunctional DD-carboxypeptidases that catalyse transfer reactions from D-alanyl-D-alanine terminated peptides. Although these PBPs are also inactivated by β-lactams, this may not be central to their antibiotic action. Nevertheless, carboxypeptidases of this type are convenient to purify and have been used widely as models for the nature of the interaction between PBPs and penicillin. The most widely studied enzymes are the extracellular DD-carboxypeptidases produced by *Streptomyces* species and carboxypeptidases solubilized from the membranes of *Escherichia coli* and *Bacillus stearothermophilus*. The *Streptomyces* enzymes display some

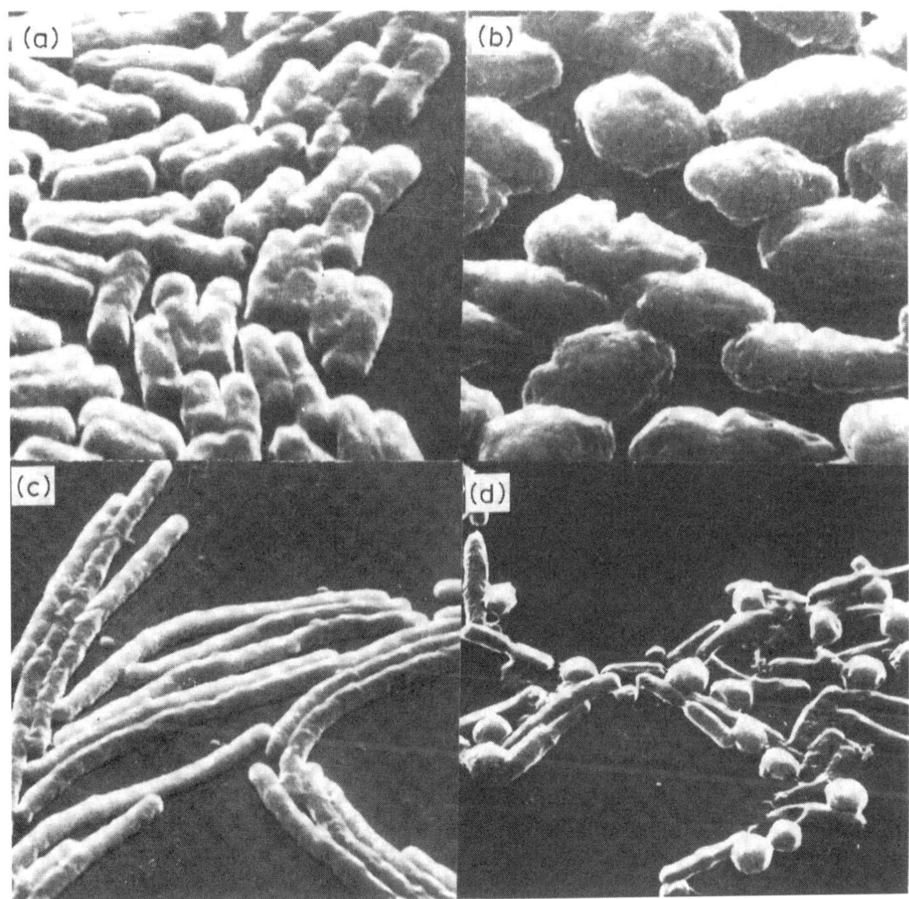

FIGURE 2.16 Effects of different β-lactam antibiotics on *Escherichia coli*, shown by scanning electron microscopy. (a) Untreated culture (×11 000). Other pictures show cells after treatment for 60 min with the antibiotics indicated: (b) mecillinam (10 μg ml^{-1}; ×12 000); (c) cephalexin (32 μg ml^{-1}; ×5800); (d) mecillinam and cephalexin combined (10 and 32 μg ml^{-1}, respectively; ×480). (Reprinted by permission of the University of Chicago Press and the authors, D. Greenwood and F. O'Grady (1973) *J. Infect. Dis.*, **128**, 793.)

transpeptidase activity besides their high carboxypeptidase activity. The interaction of a penicillin or cephalosporin (I) with the enzyme (E) can be represented as:

$$E + I \underset{k_2}{\overset{k_1}{\rightleftharpoons}} EI \xrightarrow{k_3} EI^* \xrightarrow{k_4} E + \text{degraded inhibitor.}$$

The first step is reversible binding to the enzyme. The second stage, involving chemical modification of the inhibitor with covalent binding to the enzyme, is irreversible as is the final stage of enzyme release. For high antibacterial activity k_3 should be rapid, preventing release of inhibitor through reversal of the initial binding, and k_4 should be slow to maintain the enzyme in the inactive EI* form and to avoid significant re-activation.

Measurements show that the widely used β-lactam antibiotics have just such characteristics and this scheme goes far to explain their outstanding effectiveness. There is good reason to suppose that the inactivation mechanism is the same with cross-linking transpeptidases as with DD-carboxypeptidases. The nature of the end products of penicillin degradation depends on the enzyme involved. It may be a simple opening of the β-lactam ring to give the penicilloate or there may be more extensive breakdown, leading to the production, from benzylpenicillin, of phenylacetyl glycine. Those enzymes that yield penicilloate are equivalent to slow-acting β-lactamases. There is evidence to suggest that active β-lactamases are relatives of carboxypeptidases and transpeptidases in which reaction k_4 is rapid instead of very slow.

The mechanism of action of DD-carboxypeptidases and cross-linking transpeptidases resembles that of certain esterases and amidases. These enzymes possess specially reactive groups, associated with their active centres, which undergo transient acylation in the course of enzymic action. Antibiotics containing a β-lactam ring behave chemically as acylating agents. The action of penicillin on the PBPs thus involves acylation of the enzymically active site to form the inactive complex EI*. This explanation was supported by experiments with purified DD-carboxypeptidases from *Bacillus stearothermophilus* and *Bacillus subtilis*. The enzymes were allowed to react briefly with [^{14}C]benzylpenicillin or with a substrate analogue [^{14}C]Ac_2L-Lys–D-Ala–D-lactate; D-lactic acid is the exact hydroxyl analogue of D-alanine, and use of this derivative enabled the transient enzyme reaction intermediate to be trapped. In peptide fragments from the *Bacillus stearothermophilus* enzyme, radioactivity was found in a peptide with 40 amino acid residues and the label was shown to be associated with the same specific serine residue, whether the reactant was benzylpenicillin or the substrate analogue. Similar results were found with the *Bacillus subtilis* enzyme from which a labelled 14-unit peptide was isolated. This peptide showed extensive homology with 14 residues of the *Bacillus stearothermophilus* peptide and the label was associated with the corresponding serine residue. It was concluded that penicillin binds to the active site and acylates the same serine as the substrate. Unlike the substrate, the degraded penicillin was only released very slowly (reaction k_4 in the scheme above) and thus blocked further access of substrate to the site.

How can this action of penicillin be related to its structure? The most widely quoted explanation depends on the similarity of the spatial orientation of the principal atoms and polar groups in the penicillin nucleus to one particular orientation of the D-alanyl-D-alanine end group of the pentapeptide side chain of peptidoglycan precursors (Figure 2.17). When the two structures are compared, the peptide bond between the alanine units is seen to correspond in position to the lactam group in the four-membered ring of penicillin responsible for its acylating properties. Such a group bound to the cross-linking transpeptidase, close to its active centre, could well usurp the acylating function implicit in the normal reaction of the substrate with the enzyme. When the structures (illustrated in Figure 2.17) are compared more critically, it becomes apparent that the agreement between them is imperfect, but can be much improved if the peptide bond of the D-alanyl-D-alanine end group is represented not in its normal planar form but twisted nearly 45° out of plane. This may imply that the conformation of the penicillin molecule resembles the transition state of the substrate rather than its resting form. During the enzymic transpeptidation the peptide bond quite possibly undergoes this sort of distortion. The rigidity of the bicyclic ring structures of the penicillins and cephalosporins maintains the principal binding groups in fixed relative conformation. This was thought to be an important feature in binding to the active site. However, even this concept may have to be reviewed in the light of the discovery of the monobactams.

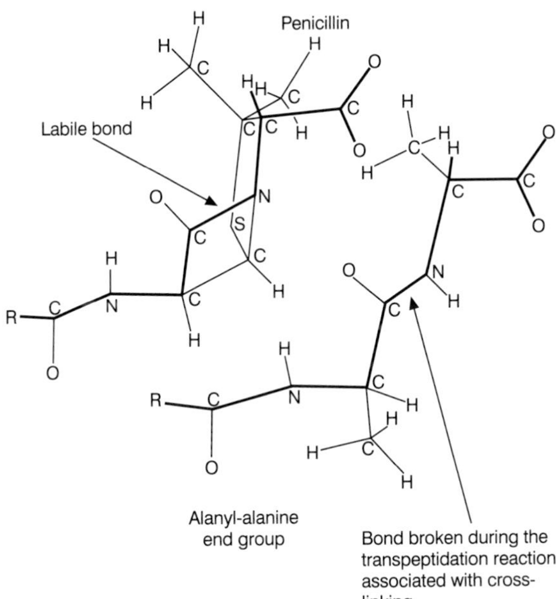

FIGURE 2.17 Comparison of the structures of penicillin with that of the D-alanyl-D-alanine end group of the peptidoglycan precursor. (Reproduced by permission of the Federation of American Societies for Experimental Biology from J. L. Strominger *et al.* (1967) *Fed. Proc.*, **26**, 18.)

2.5 Drugs that interfere with the biosynthesis of cell wall of mycobacteria

The nature of the cell envelope of mycobacteria is remarkably complex and underlies many of the characteristic properties of these organisms, including their extremely low permeability and intrinsic resistance to commonly used antibiotics. The reader is referred to a review provided in 'Further reading' at the end of this chapter for detailed information on the envelope of mycobacteria. A key feature of the mycobacterial cell envelope that distinguishes it from most other bacteria is the mycolyl-arabinogalactan–peptidoglycan complex. The long-chain mycolic acids of this complex are present mainly as attached esters of arabinogalactan and are believed to form the inner leaflet of an outer membrane. The outer leaflet is comprised of the lipid components of a range of complex glyco- and peptidolipids. Mycobacteria are notoriously impermeable to many solutes, including some antibiotics, and this low permeability is attributed to the nature of the lipid bilayer of the outer membrane.

2.5.1 Isoniazid

Isoniazid (Figure 2.18) provides one of the foundations of the combination therapy for tuberculosis. The problem of drug-resistant *Mycobacterium tuberculosis* led to the concept of combining several chemically distinct drugs with, as it later turned out, different modes of action, to minimize the risk of the emergence of resistant bacteria. In combination variously with rifampicin, ethambutol, pyrazinamide and occasionally streptomycin, isoniazid is an effective antitubercular drug which has been in use since 1952.

In the mycobacteria, the *inhA* gene encodes an enzyme that has been identified as the molecular target for isoniazid and the related drug, ethionamide. This enzyme, designated InhA, catalyses the NADH-dependent reduction of the 2-*trans*-enoyl-acyl carrier protein (ACP), an essential reaction in the elongation of fatty acids. Long-chain substrates, containing between 12 and 24 carbon atoms, are preferentially used by InhA, an observation which implicates the enzyme in the biosynthesis of the mycolic acids. Inhibition of the biosynthesis of mycolic acids therefore disrupts the assembly of the mycolyl-arabinogalactan–peptidoglycan complex and loss of cell viability. However, it is clear from studies with recombinant InhA that isoniazid itself is only a weak inhibitor of the enzyme. The compound is, in fact, first converted by oxidative cellular metabolism to a reactive metabolite which is then believed to attack NADH bound to the enzyme. Isoniazid has long been known to be metabolically unstable in mycobacteria due to the activity of a unique mycobacterial catalase–peroxidase encoded by the *katG* gene. Recent studies with the recombinant form of this enzyme show that it

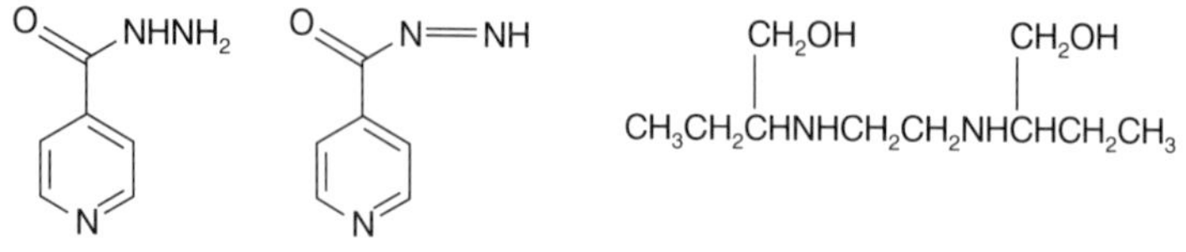

Isoniazid Isonicotinic acid Ethambutol

$CH_3CH_2CH(CH_2OH)NHCH_2CH_2NHCH(CH_2OH)CH_2CH_3$

FIGURE 2.18 Structures of synthetic compounds used in the combination therapy of tuberculosis. The structure of the microbial metabolite isonicotinic acid can be seen to resemble that of isoniazid.

converts isoniazid to several chemically reactive derivatives, isonicotinic acid (Figure 2.18) being the major product. This two-stage concept of the mechanism of action of isoniazid is strengthened by the existence of two forms of mycobacterial resistance to the drug. One type of mutant has a defective *katG* gene that precludes the conversion of the prodrug to its active form. The second resistant phenotype depends on an isoniazid-resistant variant of InhA, characterized by a markedly lower affinity for NADH which minimizes the attack of the isoniazid metabolite on the enzyme.

2.5.2 Ethambutol

The antibacterial activity of isoniazid is confined to *Mycobacterium tuberculosis*. Ethambutol, which has been in clinical use against tuberculosis since 1961, has a broader spectrum of action including *Mycobacterium avium*, a serious opportunist pathogen in patients with AIDS. Despite many years of use, the molecular basis of the bacteriostatic action of ethambutol has only been identified very recently. It had long been known that the drug in some way blocked the biosynthesis of the polysaccharide arabinan, but the actual mechanism was unknown. The target for ethambutol was eventually established by cloning the genetic elements responsible for resistance to this drug in *Mycobacterium avium*. The structural genes *embA* and *embB* encode the enzyme arabinosyl transferase III, which mediates the polymerization of arabinose into arabinan. A third gene, *embR*, encodes a transcriptional activator that upregulates the transcription of *embA* and *embB*. Resistance to ethambutol is determined by the activity of *embR* which ensures over-expression of the two structural genes for arabinosyl transferase III. The clear implication of these observations is that in ethambutol-sensitive cells the inhibition of arabinosyl transferase III underlies the antibacterial activity of the drug. There is some evidence that other arabinosyl transferases may also be affected by ethambutol but this is not thought to contribute significantly to its antibacterial action.

Disruption of the biosynthesis of the arabinogalactan component of the mycobacterial cell envelope may increase cellular permeability to other drugs. This could account for the valuable clinical synergism that is achieved when ethambutol is combined with a drug of large molecular size, such as rifampicin.

2.6 The structure of the fungal cell wall

Although it serves analogous functions to the bacterial cell wall, the structure of the fungal wall is very different from that of its bacterial counterpart. Critically, fungal walls do not contain peptidoglycan, so neither β-lactam nor glycopeptide antibiotics have any effect on the viability of fungi.

The fungal wall is a multilayered structure, the major macromolecular components of which include chitin and glucans (that confer shape and strength to the wall) and mannoproteins. Neither chitin nor glucan occurs in mammalian or bacterial cell envelopes. Mannoproteins, on the other hand, are less likely to be unique to fungi since glycosylated proteins are found in all eukaryotes. The composition and organization of the cell wall vary significantly among the various fungal species and define the identity of the organisms. Chitin is a linear β-1,4-linked homopolymer of *N*-acetylglucosamine. Glucan is a linear glucose polymer which occurs in β-1,3-, α-1,3- and β-1,6-forms, according to species. The mannoproteins comprise complex chains of mannose linearly bonded by 1,6-links to which oligomannoside side branches are attached by 1,2- and α-1,2-bonds. The polysaccharide structures are covalently linked to protein via a β-1,4-disaccharide of *N*-acetylglucosamine residues by either *N*-glycosylation of asparagine or *O*-glycosylation at the free hydroxyl groups of threonine or serine residues. Some idea of the diversity of mannoproteins may be gauged from the fact that between 40 and 60 different mannoproteins can be isolated from yeast cell walls.

The arrangement of these various polymers in the wall of the important fungal pathogen *Candida albicans* is illustrated in Figure 2.19. The insoluble polymers, chitin and glucan, confer mechanical strength on the wall. The function of the mannoproteins is less clear but appears to be essential because inhibitors of *N*- and *O*-glycosylation are lethal, although it should be remembered that the effects of inhibition of these reactions are not confined to the biosynthesis of mannoproteins.

Because the integrity of the wall is essential to the viability of fungal cells, those biochemical reactions involved in its synthesis, which are largely specific to fungi, afford excellent targets for selective antifungal agents.

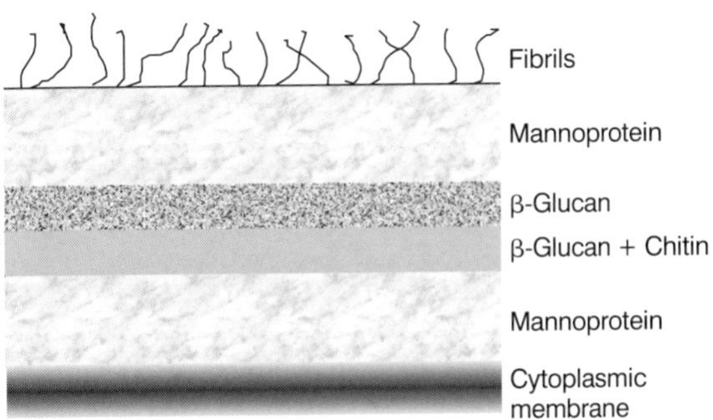

FIGURE 2.19 The general arrangement of layers in the fungal cell envelope. The components are not drawn to scale. It should be remembered that the precise structure of the fungal cell envelope is markedly species-specific.

2.7 Inhibitors of the biosynthesis of the fungal cell wall

2.7.1 Chitin

The key reaction in the biosynthesis of chitin is:

$$2n \text{ UDP-}N\text{-acetylglucosamine} \rightarrow (N\text{-acetylglucosamine-}(1,\beta)\text{-}N\text{-acetylglucosamine})_n + 2n \text{ UDP.}$$

The enzyme chitin synthase, which exists in several forms, catalyses the reaction. Two related groups of antibiotics inhibit chitin synthase, the **polyoxins** and **nikkomycins** (Figure 2.20). Both types are analogues of UDP-*N*-acetylglucosamine and presumably inhibit the enzyme by competition with this substrate. However, the existence of the isozymes of chitin synthase poses some problems for defining the activity of these antibiotics. In the yeast *Saccharomyces cerevisiae* chitin synthase 1 is required for repairing damage to the intercellular septum incurred during the separation of daughter cells. Chitin synthase 2 is specifically involved in the biosynthesis of the septum itself, while chitin synthase 3 produces most of the chitin in the bud scar and lateral cell wall. No single synthase appears to be essential for cell viability but the loss of all three in mutants of *Saccharomyces cerevisiae* is lethal. Chitin synthase exists in analogous isozymic forms in *Candida albicans* and possibly in other pathogenic fungi. The variable susceptibility of fungi to polyoxins and nikkomycins may be due to differences in the distribution and sensitivities of the isoforms to these antibiotics. Another factor which determines susceptibility to polyoxins is their transport into fungal cells by a permease that normally carries dipeptides. *Candida albicans* is intrinsically resistant to polyoxins because of the low activity of this permease. Loss of the permease seems to have little effect on cell viability. Despite these potential problems it is hoped that the best of the currently available inhibitors of chitin synthase, nikkomycin Z, may eventually find a place in clinical medicine.

2.7.2 Glucan

Echinocandin B (Figure 2.21) is a member of a large family of naturally occurring and semi-

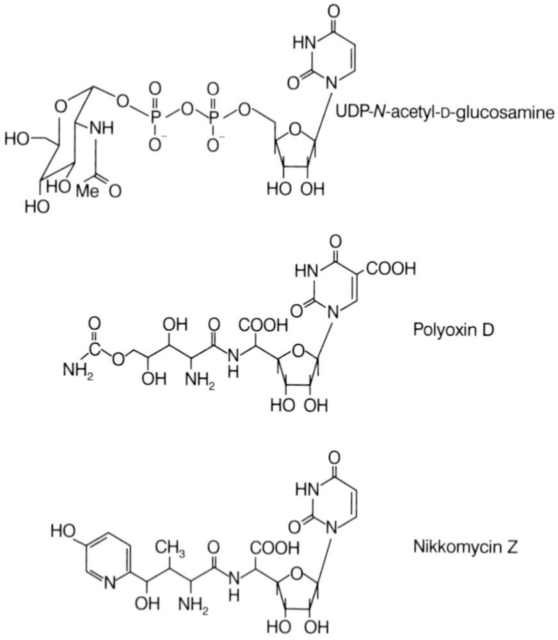

UDP-*N*-acetyl-D-glucosamine

Polyoxin D

Nikkomycin Z

FIGURE 2.20 Antifungal agents that inhibit cell wall chitin synthesis, together with the substrate UDP-*N*-acetylglucosamine.

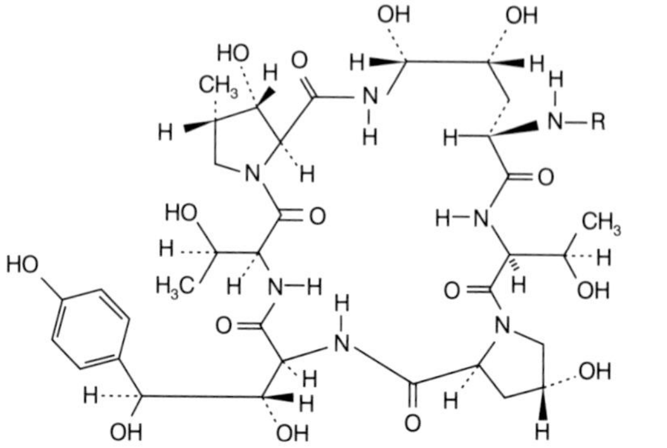

FIGURE 2.21 Echinocandin B: inhibitor of the biosynthesis of the glucan polymer in the cell wall of yeasts.

synthetically modified cyclic lipopeptides, which have potent activity against *Candida* spp. and against aspergillus. This compound, along with other members of its family, is a powerful non-competitive inhibitor of β-1,3-glucan synthase. This specificity may explain the lack of activity against fungi where glucan is not mainly β-1,3-linked. In *Cryptococcus* spp., for example, a dangerous pathogen affecting the respiratory tract, the glucan is mostly α-1,3-linked and the organisms are resistant to the cyclic lipopeptide antibiotics. The enzyme β-1,3-D-glucan synthase comprises two subunits, one of which is an integral membrane protein, molecular weight 215 kDa, with 16 transmembrane helices. The other subunit is a much smaller protein (20 kDa) that binds GTP and is only loosely associated with the cell membrane. The function of this latter protein is apparently to activate the catalytic activity of the membrane-bound protein. Studies with an echinocandin-resistant mutant of *Saccharomyces cerevisiae* have identified the membrane-bound component of β-1,3-D-glucan synthase as the target of the drug. It seems likely that this finding will extend to the major fungal pathogens.

The usefulness of echinocandin B is limited by its propensity to cause lysis of the red blood cells, apparently due to the extended lineoyl side chain. Replacement of the lineoyl group with shorter, synthetic side chains eliminates this toxicity and yields compounds with potentially useful activity against infections caused by *Candida* spp. and *Pneumocystis carinii.*

2.7.3 Mannoproteins

Pradimicin A (Figure 2.22) belongs to a unique group of antibiotics originally isolated from *Actinomadura hibisca* and is active against *Candida* spp., *Cryptococcus* spp. and *Aspergillus* spp. The antifungal action involves a change in the permeability of the cell membrane, which may result from the ability of pradimicin to form an insoluble complex with mannan in the presence of calcium ions. Although this points to some form of interference with mannoprotein function the biochemistry of the antifungal action of pradimicin requires further investigation.

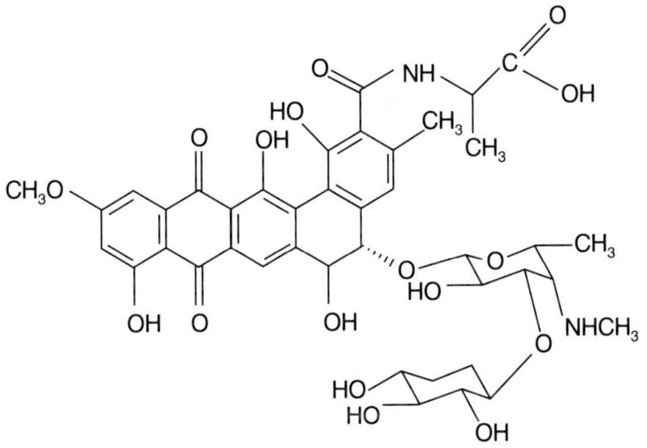

FIGURE 2.22 Pradimicin A, an antibiotic active against several species of yeast pathogens. Its mode of action may depend upon interference with the function of cell wall mannoproteins.

Further reading

Brennan, P.J. (1995). The envelope of mycobacteria. *Ann. Rev. Biochem.* **64**, 29-63.

Debono, M. and Gordee, R.S. (1994). Antibiotics that inhibit fungal cell wall development. *Ann. Rev. Microbiol.* **48**, 471-97.

Doyle, R.J. and Marquis, R.E. (1994). Elastic, flexible peptidoglycan and bacterial cell wall properties. *Trends Microbiol.* **2**, 57-60.

Georgopapadakou, N.H. and Tkacz, J.S. (1995). The fungal cell wall as a drug target. *Trends Microbiol.* **3**, 98.

Ghuysen, J.-M. (1991). Serine β-lactamases and penicillin-binding proteins. *Ann. Rev. Microbiol.* **45**, 37-67.

Ghuysen, J.-M. *et al.* (1996). Penicillin and beyond: evolution, protein fold, multimodular polypeptides and multiprotein complexes. *Microb. Drug Resist.* **2**, 163-75.

Hancock, R.E. (1997). The bacterial outer membrane as a drug permeability barrier. *Trends Microbiol.* **5**, 37-42.

Johnsson, K. and Schulz, P.G. (1994). Mechanistic studies of the oxidation of isoniazid by the catalase-peroxidase from *Mycobacterium tuberculosis. J. Am. Chem. Soc.* **116**, 7425-6.

Kurtz, M.B. and Douglas, C.M. (1997). Lipopeptide inhibitors of fungal glucan synthase. *J. Med. Vet. Mycol.* **35**, 79.

Prescott, L.M., Harley, J.P. and Klein, D.A. (1996). *Microbiology,* Wm. C. Brown, Dubuque IA.

Schäffer, L.M., Schneider, T.R and Sheldrick G.M. (1996). Crystal structure of vancomycin. *Structure* **4**, 1509-15.

Skarzynski, T. *et al.* (1996). Structure of UDP-*N*-acetyl glucosamine enol pyruvyl transferase: an enzyme essential for the synthesis of bacterial peptidoglycan, complexed with substrate UDP-*N*-acetyl glucosamine and the drug fosfomycin. *Structure* **4**, 1465-74.

Chapter three

Antiseptics, antibiotics and the cell membrane

3.1 Microbe killers: antiseptics and disinfectants

The major interest throughout this book lies in the mechanism of action of drugs that can be used against microbial infections. For this purpose the compound must normally be absorbed and circulate in the blood. However, there is also a requirement in medicine and in industry for substances that kill bacteria and other micro-organisms on the surface of the body or in other places. Such products are known as disinfectants, sterilants, antiseptics or biocides, the choice of term depending on the circumstances in which they are used. 'Disinfectant' describes products intended for use in the presence of dirt and dense bacterial populations, for example in the cleaning of animal quarters or drains. 'Biocide' is used more particularly for preservatives that prevent bacterial and fungal attack on wood, paper, textiles and other kinds of organic material and also in pharmaceutical preparations. 'Antiseptic' is a term usually reserved for a substance that can be safely applied to the skin and mucosal surfaces with the aim of reducing the chances of infection by killing the surface bacteria. 'Sterilants' are substances used to sterilize an enclosed space; since penetration is paramount in this application, sterilants are usually gaseous.

The requirements for a compound having disinfectant or antiseptic action differ markedly from those needed in a systemic drug. Many compounds used successfully against microbial infections do not actually kill micro-organisms, but only prevent their multiplication; most are inactive against non-growing organisms. A cessation of microbial growth is often all that is needed in treating an infection, since the body has antibody and phagocytic defences that can soon be mobilized to remove pathogens present in relatively small numbers. Furthermore, systemic antimicrobial agents often have a fairly limited spectrum of action. This is acceptable since the compound can be selected according to the nature of the infection that is being treated. Antiseptics, by contrast, are usually required to have a broad-spectrum, killing effect. Antiseptics and preservatives used in ointments, creams, eye-drops and multi-dose injections must obviously be free from toxicity against the host tissues.

A distinction is often made between 'static' and 'cidal' compounds, but the division is by no means clear-cut. There is no certain way of determining whether a micro-organism is dead. The usual method of assessing the killing effect of an antiseptic is by measuring the 'viable count' of a previously treated bacterial or fungal suspension. The antiseptic is first inactivated and dilutions of the suspension are added to a rich medium. The organisms are deemed to be alive if they give rise to colonies. Many compounds are static at low concentrations and cidal at higher concentrations, and the effect may also depend on the conditions of culture. However, for antiseptics and disinfectants a cidal effect is required under all normal conditions of application. Such compounds must be able to kill micro-organisms whether they are growing or resting and must be able to deal with most of the common bacteria likely to be found in the environment and, ideally, fungi and viruses as well. Bacterial and fungal spores are usually much more difficult to kill.

Many of the older disinfectants are compounds of considerable chemical reactivity. Their antimicrobial action presumably depends on their ability to react chemically with various groups on or in the organism, thus killing it. Such compounds include hydrogen peroxide, the halogens and hypochlorites, the gaseous sterilants ethylene oxide, ozone, etc. Salts and other derivatives of the heavy metals, particularly of mercury, probably owe their antimicrobial effect to reaction with vital thiol groups. In disinfection, their high reactivity and toxicity limit their scope and they are not now generally acceptable for the more delicate uses as antiseptics. For this purpose two main groups of compounds are used almost exclusively, the phenols and the cationic antiseptics. The main emphasis with these agents has been their efficacy against bacteria. Increasingly, however, there is concern that they should have useful activity against fungi and viruses. Although there are differences between the actions of phenols and cationic antiseptics, they show many common features (their effects have been most closely investigated in bacteria):

1. Antiseptics bind readily to bacteria, the amount adsorbed increasing with an increasing concentration in solution. The adsorption isotherm sometimes shows a point of inflection which corresponds to the minimum bactericidal concentration; higher concentrations lead to a much greater adsorption of the compound. The most important site of adsorption is the cytoplasmic membrane. Spheroplasts or protoplasts lacking the outer cell wall layers will bind the antiseptic and may be lysed or damaged. Adsorption by isolated cell membranes can also be demonstrated.
2. The extent of killing of bacteria is governed by three principal factors:

   (a) concentration of the antiseptic;
   (b) bacterial cell density; and
   (c) time of contact.

   The adsorption of a given amount of the compound per cell leads to the killing of a definite fraction of the bacterial population in a chosen time interval.
3. The lowest concentration of the antiseptic that causes death of bacteria also brings about leakage of cytoplasmic constituents of low molecular weight. The most immediately observed effect is a loss of potassium ions. Leakage of nucleotides is often detected by the appearance in the medium of material having an optical absorption maximum at 260 nm. Gram-positive cells show leakage of amino acids. Some loss of cytoplasmic solutes is not in itself lethal. Compounds are known that cause this effect but do not kill bacteria; moreover bacteria that have been rendered leaky by low concentrations of an antiseptic will often grow normally if they are immediately washed and placed in a nutrient medium. The increased permeability is a sign of changes in the membrane which are initially reversible but become irreversible on prolonged treatment.
4. The necessary characteristic of antiseptics is their bactericidal action, but there is often a low and rather narrow concentration range in which their effect is bacteriostatic. At these low concentrations certain biochemical functions associated with the bacterial membrane may be inhibited.
5. In the presence of higher concentrations of antiseptic and after prolonged treatment, the compound usually penetrates the cell and brings about extensive disruption of normal cellular functions, including precipitation of intracellular proteins and nucleic acids.

The primary effect of these antiseptics on the cytoplasmic membrane is thus established beyond doubt, but secondary actions on cytoplasmic processes are less well defined and may vary from one compound to another. Examples of evidence of action for particular compounds will be given as illustrations.

3.1.1 Phenols

Crude mixtures of cresols solubilized by soap or alkali and originally introduced as 'lysol' are still used as rough disinfectants. They need to be applied at high concentrations and are irritant and toxic but they kill bacteria, fungi and some viruses. For more refined applications as antiseptics, chlorinated cresols or xylenols are commonly used since they are less toxic than the simpler phenols. In general, the primary action of the phenolic disinfectants and antiseptics is to cause the denaturation of microbial proteins, the first target being the proteins of the cell envelope, leading to lethal changes in membrane permeability.

3.1.2 Alcohols

Alcohols are still used as cheap disinfectants and preservatives. Ethanol, for example, is a reasonably effective skin disinfectant as a 60–70% solution which kills both bacteria and viruses. Isopropanol (propan-2-ol) (at least 70%), which is slightly more effective as a bactericide than ethanol but is more toxic, can be used to sterilize instruments such as clinical thermometers. The more complex compound known as bronopol (2-bromo-2-nitropropane-l,3-diol) is an effective preservative for certain pharmaceutical products and toiletries, although there are concerns about the potential toxicity of some of the decomposition products of this compound upon its exposure to light.

The antibacterial effects of the alcohols can be traced to a disruption of membrane function. The action of short-chain alcohols such as ethanol is probably dominated by the polar function of the hydroxyl group which may form a hydrogen bond with the ester groups of membrane fatty acid residues. In contrast with ethanol, longer-chain alcohols gain access to the hydrophobic regions of membranes and this probably accounts for the increasing potency of antimicrobial action up to a maximum chain length of 10 carbon atoms. The interaction of alcohols with cell membranes produces a generalized increase in permeability which is lethal at higher concentrations. Bronopol may exert an additional effect by interacting with thiol groups in membrane proteins.

3.1.3 Cationic antiseptics

This classification covers a number of compounds differing considerably in chemical type. Their common features are the presence of strongly basic groups attached to a fairly massive lipophilic molecule. Although antiseptic action is found quite widely in compounds having these characteristics the degree of activity is sharply dependent on structure within any particular group. For instance in cetrimide (Figure 3.1) the length of the main alkyl chain is 14 carbon atoms and the activity of other compounds in the same series falls off markedly with longer or shorter chains. Cetrimide combines excellent detergent properties and minimal toxicity with a useful antiseptic action. However, it is not very potent against *Proteus* and *Pseudomonas* species and has little antiviral activity, except against viruses with a lipid envelope. Experiments with *Escherichia coli* labelled with ^{32}P have shown that with increasing concentrations of cetrimide the loss of cell viability closely parallels the degree of leakage of radioactivity from the bacteria. An effect on bacterial growth, however, is noticeable at concentrations that affect neither viability nor permeability.

One of the best and most widely used of the cationic antiseptics is chlorhexidine (Figure 3.1). This compound has two strongly basic groups, both biguanides; it is often formulated as the digluconate which has good solubility in water. Chlorhexidine is much less surface active than cetrimide and has little detergent action. However, it acts against a wide range of bacteria at concentrations between 10 and 50 $\mu g\ ml^{-1}$ and it also has

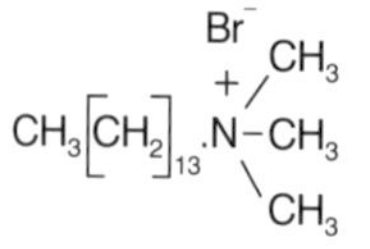

$CH_3[CH_2]_{13}.\overset{+}{N}(CH_3)_3\ Br^-$

Cetrimide

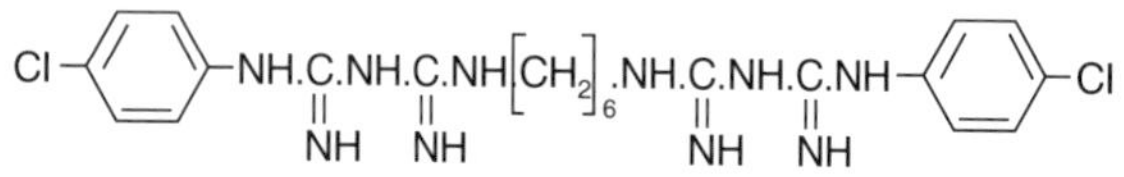

$Cl-C_6H_4-NH.\underset{\|}{C}(NH).NH.\underset{\|}{C}(NH).NH[CH_2]_6.NH.\underset{\|}{C}(NH).NH.\underset{\|}{C}(NH).NH-C_6H_4-Cl$

Chlorhexidine

FIGURE 3.1 Synthetic antiseptics. The formula for cetrimide shows the main components in the preparations normally sold. Homologues with other chain lengths, especially C_{16}, are also present.

useful activity against *Candida albicans*. Its toxicity is low and it has so little irritancy that it can be used on the most sensitive mucosal surfaces. For example, it is a useful aid to oral hygiene. Periodic rinsing of the mouth with chlorhexidine solution greatly reduces the population of *Streptococcus mutans* on the teeth. This minimizes the production of dental plaque and reduces periodontal infections that give rise to gingivitis. It also decreases the incidence of some types of caries. An important feature of this action is the strong binding of chlorhexidine to the tissues in the mouth, including the teeth, with subsequent slow release which maintains an antibacterial action over an extended period.

Chlorhexidine exerts effects on the cytoplasmic membrane characteristic of cationic antiseptics. At concentrations that just prevent growth of *Streptococcus faecalis* it inhibits the adenosine triphosphatase of the membrane. The effect can be shown in isolated membranes and on the solubilized enzyme derived from them. A similar concentration of chlorhexidine inhibits the net uptake of potassium ions by intact cells, and the two effects are thought to be interdependent. This interaction between chlorhexidine and the cell membrane probably involves electrostatic binding between the cationic groups of the antiseptic and anionic residues in the membrane. Hydrophobic interactions between the hexamethylene chains of chlorhexidine and the aliphatic chains of membrane lipids also contribute to the stability of the complex. When bacteria are treated with a range of concentrations of chlorhexidine and then examined for leakage of cytoplasmic solutes, the degree of leakage increases with concentration up to a maximum and then declines at higher concentrations. Low concentrations of chlorhexidine provoke the release of K^+ ions, nucleotides and sugars. Electron microscopy shows that the cells from higher levels of chlorhexidine treatment are grossly altered. The increased membrane permeability apparently allows the antiseptic to enter the cytoplasm and to cause precipitation of the nucleic acids and proteins, resulting in the death of the cells. Under these circumstances leakage is probably prevented by simple mechanical blockage.

With Gram-negative bacteria chlorhexidine damages the outer membrane as well as the cytoplasmic membrane. This can be seen as 'blistering' in electron micrographs (Figure 3.2). This phenomenon will be discussed further in connection with the action of polymyxin.

3.2 Polypeptide antibiotics

Several classes of polypeptide antibiotics are known. In two groups the effects on bacteria conform exactly with the properties already discussed for the phenolic and cationic antiseptics. They owe their primary antibacterial action to their binding to the cytoplasmic membrane, with subsequent disturbance of its function, and can therefore be regarded as cytolytic agents. Both types are cyclic polypeptides. One group includes the tyrocidins and gramicidin S which are cyclic decapeptides (Figure 3.3). These contain one or sometimes two free amino groups. They are more active against Gram-positive than against Gram-negative bacteria. The polymyxins which form the second group have a smaller polypeptide ring attached to a polypeptide chain terminating with a branched 8- or 9-carbon fatty acid residue. They have five free

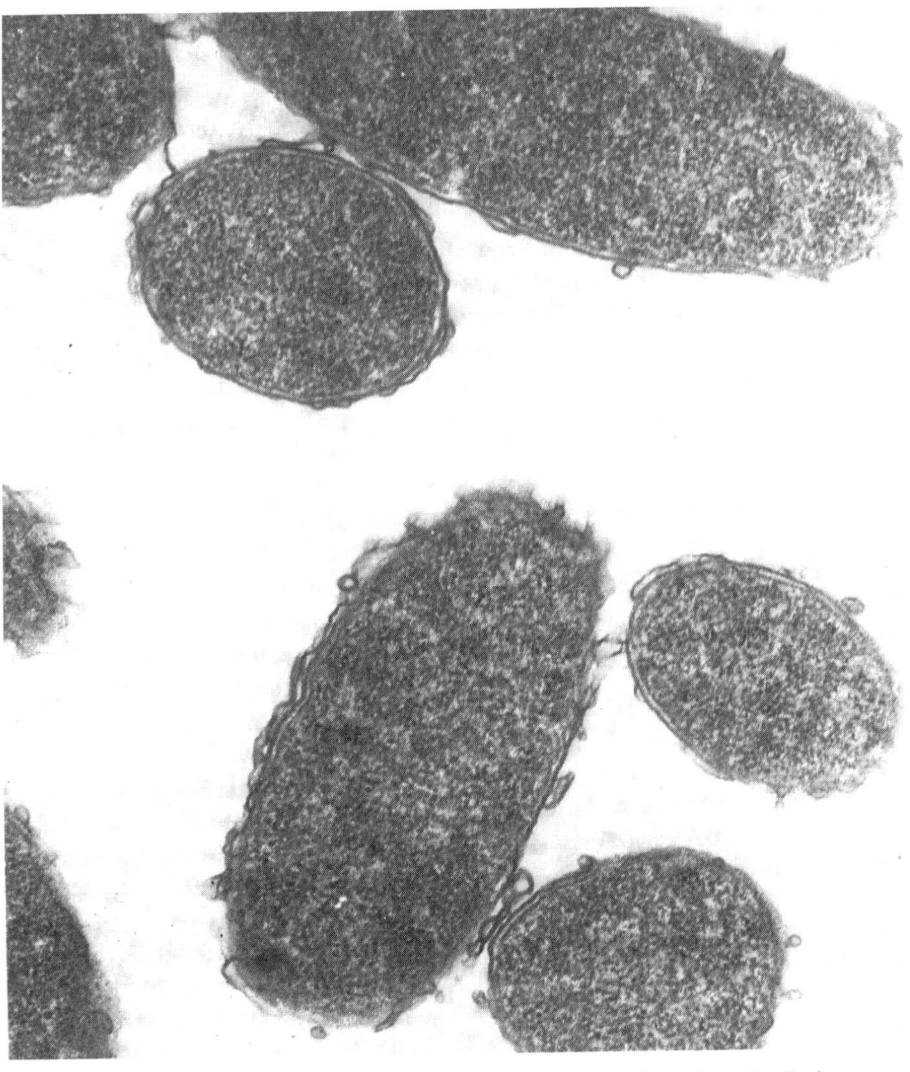

FIGURE 3.2 Electron micrograph of a cross-section of an *Escherichia coli* cell after treatment with chlorhexidine (30 μg ml^{-1}), showing 'blistering' of the cell envelope.

amino groups associated with the diaminobutyric acid units (Figure 3.3). The octapeptins have a similar ring structure but differ in the composition of the side chain. The antibacterial action of polymyxins and octapeptins is directed particularly against Gram-negative organisms. This selectivity can be altered dramatically by chemical modification. Thus, the penta-*N*-benzyl derivative is highly active against Gram-positive bacteria.

Polypeptide antibiotics have only a minor place in medicine because they also damage mammalian cell membranes. The polymyxins may be used systemically in severe *Pseudomonas* infections, though there is considerable risk of kidney damage. Polymyxin is bactericidal and acts on non-growing as well as on growing cells. At low concentrations its bactericidal action parallels the degree of release of cytoplasmic solutes. It is strongly and rapidly bound

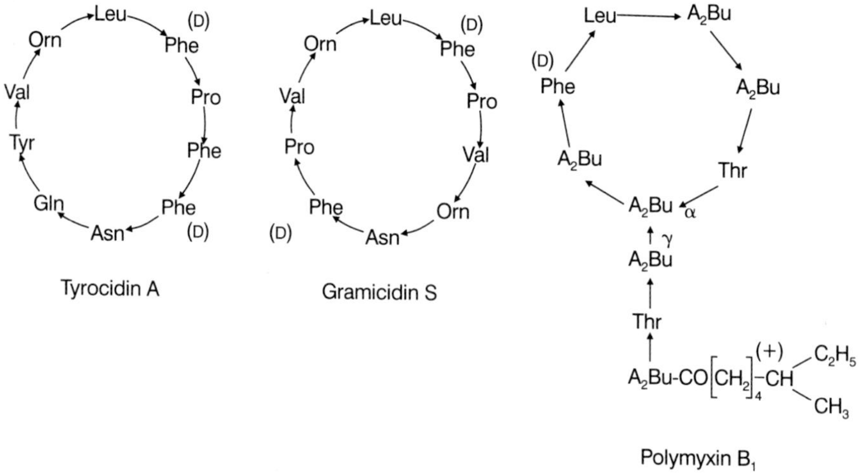

Tyrocidin A

Gramicidin S

Polymyxin B_1

FIGURE 3.3 Antibiotics that damage bacterial cell membranes. A_2Bu: 2,4-diaminobutyric acid. Arrows show direction of the peptide bonds. Except where shown, all peptide linkages involve α-amino and α-carboxyl groups. Configurations are L unless otherwise indicated.

to bacteria. With *Salmonella typhimurium* the binding of 2×10^5 molecules of polymyxin per cell was shown to be bactericidal. In Gram-negative bacteria, antibiotics of the polymyxin group apparently bind first to the outer membrane, affecting mainly the lipopolysaccharide. The gross effects of polymyxin on the outer membrane are sometimes revealed in electron micrographs as blisters, similar to those caused by chlorhexidine (Figure 3.2). The swellings may be due to an increase in the surface area of the outer leaf of the outer membrane. The parallels between the action of polymyxin and chlorhexidine are quite striking. In both, the binding and antibacterial effects are antagonized by excess of calcium or magnesium ions, indicating that the displacement of divalent ions is an important feature of their action. The disorganization of the outer membrane by polymyxin enables the antibiotic to gain access to the cytoplasmic membrane which, in turn, is damaged by the antibiotic. The disruption to normal membrane function brought about by polymyxin results in a generalized increase in membrane permeability and the loss of essential cellular nutrients and ions, such as K^+.

Physical measurements of various kinds all support the conclusion that the antibacterial action of polymyxin is caused primarily by its binding to membranes. The positively charged peptide ring is thought to bind electrostatically to the anionic phosphate head groups of the membrane phospholipid, displacing magnesium ions which normally contribute to membrane stability. At the same time the fatty acid side chain is inserted into the hydrophobic inner region of the membrane. The effect is to disturb the normal organization of the membrane and to alter its permeability characteristics.

The tyrocidins are also bactericidal and promote leakage of cytoplasmic solutes. Their action on the bacterial membrane permits passage into the cell of ions that are normally excluded, and under some conditions this causes uncoupling of oxidative phosphorylation as a secondary effect. Gramicidin S, a closely related compound, acts similarly. It lyses protoplasts from *Micrococcus lysodeikticus* but not those from *Bacillus brevis*. Since it is bactericidal towards the former organism but not the latter, it is reasonable to suppose that both its action and specificity depend upon its effect on the cytoplasmic membrane. The tyrocidins act not only on bacteria but also on the fungus *Neurospora crassa.* In this organism concentrations of the antibiotic that stop growth and cause leakage of cell contents also cause an immediate fall in membrane potential, a consequence of the destruction of the permeability barrier.

In both the tyrocidin group and in the polymyxins the cyclic molecular structure is important for antibacterial activity. The presence of basic groups is also essential, but in other respects the molecules can be varied considerably without losing activity. The simple symmetrical structure of gramicidin S has been subjected to many modifications. Activity is preserved when the ornithine units are replaced by arginine or lysine groups but is lost by modifications destroying the basic character of the terminal groups. The compound in which glycine replaces L-proline is fully active. Moreover one L-proline residue together with the adjacent D-phenylalanine can be replaced by a δ-aminopentanoic acid group without losing antibacterial activity. The resulting compound has only nine peptide groups, but retains the same ring size. Acyclic compounds having the same sequence of amino acids as gramicidin S show only slight antibacterial action.

The importance of the cyclic structure lies in the maintenance of a well-defined, compact conformation in solution. This has been shown by nuclear magnetic resonance, optical rotatory dispersion and other physical measurements. In tyrocidin A and gramicidin S the conformation is determined by lipophilic association between the non-polar side chains of the amino acids, particularly leucine, valine, proline and phenylalanine, and by hydrogen bonding between the peptide groups. Three regions have been defined in the molecular topography of tyrocidin A: a hydrophobic surface; a flat hydrophilic opposite surface, consisting of the peptide groups of most of the amino acids in equatorial positions; and a helical hydrophilic region, accommodating the amide groups of asparagine and glutamine and the tyrosine hydroxyl group. Gramicidin S shows a similar arrangement, based on a pleated-sheet structure. In both antibiotics the ornithine amino groups, which are essential for antibacterial activity, stand out from the hydrophilic surface.

3.3 Ionophoric antibiotics

Several classes of antibiotics may be grouped together because of their common property of facilitating the passage of inorganic cations across membranes by the formation of hydrophobic complexes with the ions or by forming ion-permeable pores across the membranes. Although these compounds were discovered through their antibacterial activity, they are not used in human bacterial infections because of their lack of specificity. They act equally effectively on the membranes of animal cells and may therefore be toxic. However, some ionophores have applications in veterinary medicine and animal husbandry which are discussed later. Ionophoric compounds are also of considerable biochemical interest and are widely used as experimental tools. As antimicrobial agents they are active against Gram-positive bacteria while Gram-negative bacteria are relatively insensitive because their outer membranes are impermeable to hydrophobic compounds of the molecular size of the ionophores. Monensin and lasalocid have useful activity against the protozoal parasite *Eimeria tenella,* the organism that causes coccidiosis in poultry. There is also some recent evidence for the selective toxicity of certain ionophores against the malarial parasite (see below).

3.3.1 Valinomycin

This was the first member of a group of related compounds to be discovered and is among the most widely studied. It is a cyclic depsipeptide in which amino acids alternate with hydroxy acids in a ring that contains both peptide and ester groups (Figure 3.4). An important feature, which is common to all the cyclic ionophores, is the alternation of D- and L-configurations in pairs around the 12 components of the ring structure. Valinomycin forms a well-defined complex with potassium ions. X-ray analysis of this complex reveals a highly ordered structure (Figure 3.5) in which the potassium atom is surrounded by six oxygen atoms. The ring structure is puckered and held in a cylindrical or bracelet-like form by hydrogen bonds roughly parallel to its axis. The ability to achieve such a conformation depends entirely on the alternation of D- and L-centres. The dimensions are such that the potassium atom is exactly accommodated. The ion entering the complex must shed its normal hydration shell; the complex retains the positive charge carried by the ion. The structure observed in the crystal is substantially maintained in solution. Although valinomycin will also form a complex with sodium, the smaller sodium atom fits much less exactly into the structure and this complex has a stability constant 1000 times smaller than that of the potassium complex.

The high specificity of valinomycin towards the potassium ion and the physical properties of the complex are consistent with its postulated action on biological membranes. The binding of the potassium ion in the structure of valinomycin increases the lipophilicity of the antibiotic and thereby promotes its diffusion into the hydrophobic regions of the membrane. The lipophilic molecule moves physically through the membrane lipids, carrying potassium, and returns in the protonated form. In a passive membrane the flow is determined solely by the concentration of potassium ions on each side of the membrane, but in mitochondria supplied with an energy source potassium is taken in by an energy-coupled process against the concentration gradient. The process is highly effective, one valinomycin molecule being able to transport 10^4 ions per second, a turnover rate higher than that of many enzymes. The transport of potassium by valinomycin and similar ionophores shows saturation kinetics with respect to the cation; sodium ions inhibit potassium transport although they undergo little transport themselves. The kinetic results are well explained by a model in which the ionophore at the membrane surface first forms a hydrophilic cation complex. This is transformed to a hydrophobic complex which can then cross the membrane. The rate of the transformation from one type of complex to the other determines the turnover number.

When *Streptococcus faecalis* grows anaerobically there is no oxidative phosphorylation mechanism and ATP is generated solely by glycolysis. Valinomycin inhibits this organism in normal media of low potassium content. It specifically drains the cell of potassium and growth ceases

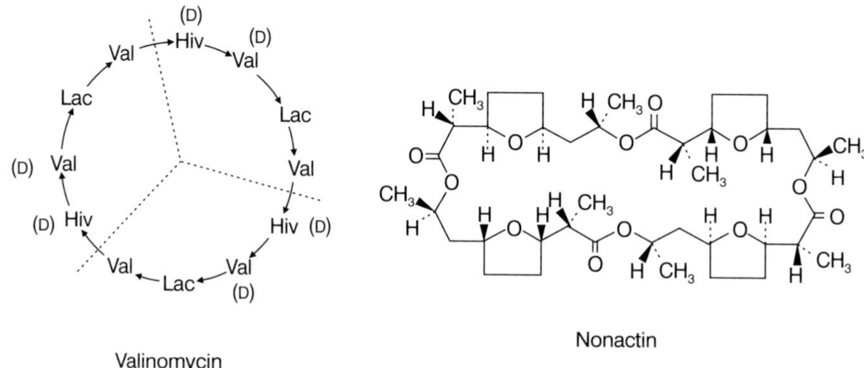

FIGURE 3.4 Antibiotics enhancing the permeability of membranes to potassium ions. In the valinomycin structure: Val represents valine; Lac, lactic acid; and Hiv, 2-hydroxyisovaleric acid. Arrows indicate the direction of peptide or ester bonds. Configurations are L unless otherwise indicated. Dotted lines separate the repeating units.

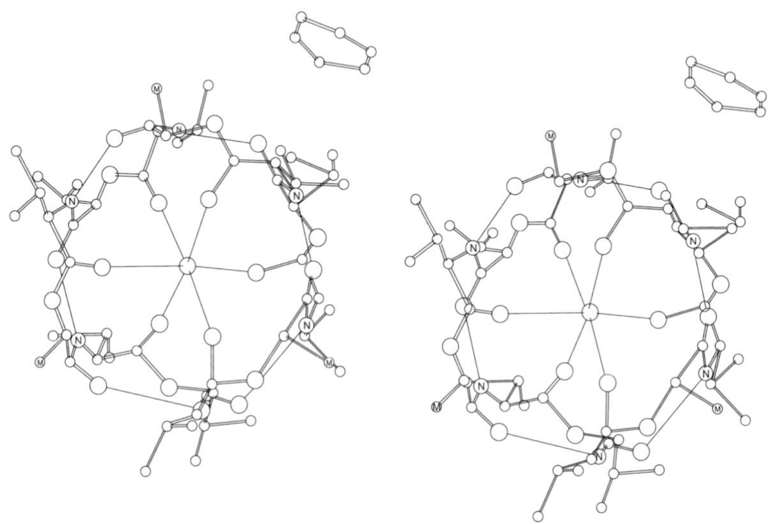

FIGURE 3.5 Stereophotographs of a model of the potassium complex of valinomycin. To obtain a three-dimensional effect the diagram should be held about 50 cm from the eyes and attention concentrated on the space between the two pictures. With practice three pictures can be seen, the middle one showing a full stereoscopic effect. The central metal ion is seen co-ordinated to six oxygen atoms. Nitrogen atoms are labelled N and the methyl groups of the lactyl residues M. Hydrogen bonds are shown by thin lines. The solitary hexagonal ring is hexane of crystallization. We are grateful to Mary Pinkerton and L. K. Steinrauf for allowing us to reproduce this picture.

because of the requirement for potassium in cellular metabolism. If the potassium content of the medium is raised to that normally present in the cytoplasm, the inhibitory action of valinomycin is prevented. With growing aerobic bacteria, the inhibitory action of valinomycin is also a function of its potassium-transporting effect, which disturbs oxidative phosphorylation as a secondary consequence. In analogous fashion, valinomycin disrupts oxidative phosphorylation in the mitochondria of eukaryotic cells.

3.3.2 Nonactin

Another series of antibiotics known as the macrotetrolides, exemplified by nonactin (Figure 3.4), have a cyclic structure which similarly permits the enclosure of a potassium ion in a cage of eight oxygen atoms (the carbonyl and tetrahydrofuran oxygens), with the rest of the molecule forming an outer lipophilic shell. To produce this structure the ligand is folded in a form resembling the seam of a tennis ball and is held in shape by hydrogen bonding. The action of the macrotetrolides closely resembles that of valinomycin.

3.3.3 Monensin

Ionophoric antibiotics of another broad group, typified by monensin (Figure 3.6) and the closely related nigericin, carry a carboxyl group. In these compounds the molecule itself is not cyclic, but as with valinomycin a metal complex is formed in which the ion is surrounded by ether oxygen atoms and the outer surface is lipophilic. This involves a folding of the molecule which brings the carboxyl group at one end into a position where it can form strong hydrogen bonds with the alcohol groups at the other end; the structure is thus stabilized into an effectively cyclic form. Monensin binds sodium ions in preference to potassium ions, but with the closely related compound nigericin, which allows a slightly wider spacing of the oxygen atoms, the selectivity is reversed. The presence of a carboxyl group in these compounds makes an important difference to their action. They promote electrically neutral cation–proton exchange across the membrane by moving as an undissociated acid in one direction and as a cation–anion complex with no net charge in the other direction. This distinguishes them from valinomycin and nonactin where the metal complex carries a positive charge.

Monensin is a compound of considerable commercial importance. It was first introduced as a treatment for the protozoal infection cocciodiosis in chickens, and proved to be of exceptional utility. It has shown few signs of the development of resistance which usually terminates the effective life of drugs sold for treating coccidiosis. Monensin also improves the utilization of feedstuffs in ruminants. Its action depends on altering the balance of free fatty acid production by rumen bacteria in favour of propionate at the expense of acetate. Propionate is energetically more useful to the animal than acetate. There is also a lessening in the metabolically wasteful production of methane. The molecular basis of these effects is uncertain but the shifts in rumen metabolism can probably be attributed to differential antimicrobial actions on the complex population of micro-organisms in the rumen. The action of monensin on cell membranes is not species specific. Its lack of toxicity when given orally to farm animals probably depends upon its limited absorption from the gastrointestinal tract.

The ionophores considered so far form complexes only with monovalent metal ions. A few ionophores are known that form complexes with divalent ions. One of the best known of these is the antibiotic calcimycin, otherwise known as A23187 (Figure 3.6). This forms a 2:1 complex with cal-

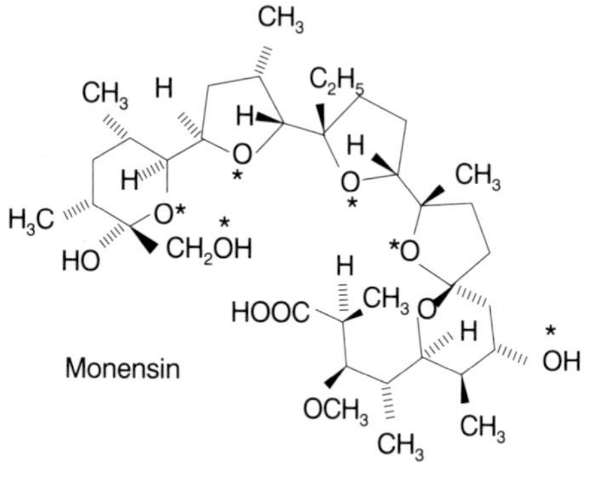

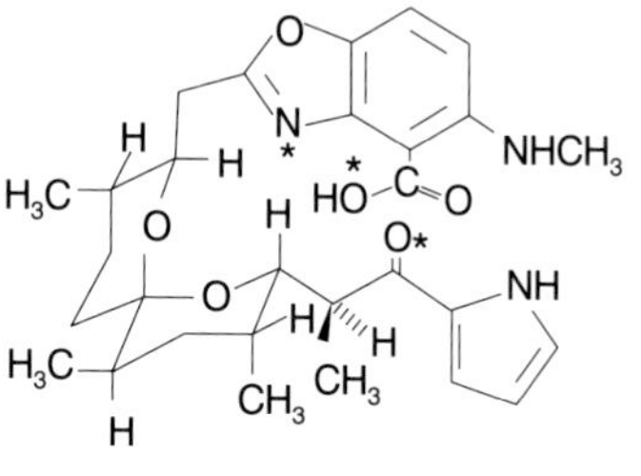

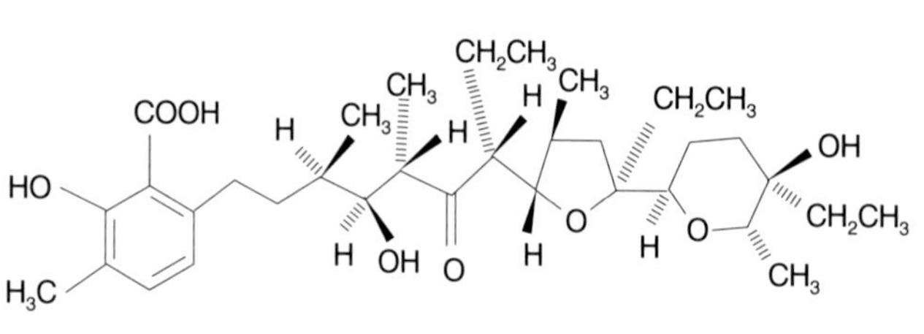

FIGURE 3.6 Three more ionophoric antibiotics. Monensin preferentially complexes sodium ions by co-ordination with the oxygen atoms marked by asterisks. The other two compounds complex with calcium ions. All three compounds have coccidiostatic activity.

cium or magnesium ions, the calcium complex having the higher stability; it binds monovalent ions only weakly. As with monensin it is not a cyclic molecule but is able to fold into an effectively cyclic conformation by the formation of a hydrogen bond between a carboxyl oxygen and the NH group of the pyrrole ring. The divalent metal ion is held in octahedral co-ordination between the polar faces of two ligand molecules. This gives an electrically neutral complex with a hydrophobic outer surface. It acts as a freely mobile carrier of these ions and causes progressive release of magnesium, uncoupling of oxidative phosphorylation and inhibition of adenosine triphosphatase in mitochondria suspended in a magnesium-free medium. Like monensin, divalent cationophoric antibiotics have coccidiostatic activity, for example the antibiotic lasalocid, or X-537A (Figure 3.6).

3.3.4 Gramicidin A

Gramicidin A (Figure 3.7) (quite unrelated to gramicidin S) has many biochemical properties resembling those of valinomycin. It shows a specificity towards potassium ions and promotes their passage across lipid membranes. However, studies have shown that its mechanism of action is different. The most significant demonstration of this distinction depends upon measurements of the electrical conductivity of artificial membranes separating aqueous layers containing potassium ions. Conditions can be chosen where addition of valinomycin, nonactin or gramicidin A at 0.1 μM concentration lowers the resistance of the membrane at least 1000-fold. If the temperature is now lowered gradually, the membrane reaches a transition point at which its lipid layer effectively changes phase from liquid to solid. In the presence of valinomycin or nonactin a 2 °C fall in temperature at the transition point causes a dramatic rise in membrane resistance, but in a similar experiment with gramicidin A resistance rises only slowly as the temperature falls. The effect with compounds of the valinomycin type is understandable since they require a liquid membrane for mobility and movement. Gramicidin A acts by the formation of a pore that permits the flow of ions through a rigid membrane. Gramicidin A is a linear polypeptide in which alternating amino acid residues have the L-configuration. The remaining residues are either D-amino acids or glycine. The C-terminus is amidated with ethanolamine and the N-terminus carries a formyl group. The configuration allows the molecule to form an open helical structure held together by hydrogen bonds lying almost parallel to the axis of the cylinder. One possible helical form is shown in Figure 3.7. The inside of the helix is lined with polar groups and there is a central hole about 0.4 nm in diameter. The fatty side chains of the amino acids form a lipophilic shell on the outside. One such molecule is not long enough to form a pore across a membrane, but head to head dimerization is believed to occur by bonds between the formyl groups. The existence of dimerization is supported by measurements in artificial membranes which show that conductance is proportional to the square of the concentration of gramicidin A. The length of the dimer is calculated to be 2.5–3.0 nm, which is somewhat less than the thickness of the fatty acid layer in many membranes so some distortion probably occurs during pore formation.

Conductivity measurements suggest that these pores have a transient existence, a small fraction of the antibiotic being in the form of pores at any given time. The life of a channel measured in a phosphatidylethanolamine artificial membrane was only 0.35 s. However, while a pore is in existence, its transporting capacity is high. One channel is estimated to convey 3×10^7 K^+ ions s^{-1} under a potential gradient of 100 mV. Thus a low concentration of gramicidin A is a very effective carrier of potassium ions. Divalent cations are too large to traverse the gramicidin pores but block the free passage of monovalent ions.

Antiprotozoal activity of ionophores

With the exception of monensin and lasalocid, the practical applications of ionophores are very limited. Recently, however, the pore-forming ionophores gramicidin D, a linear peptide structurally

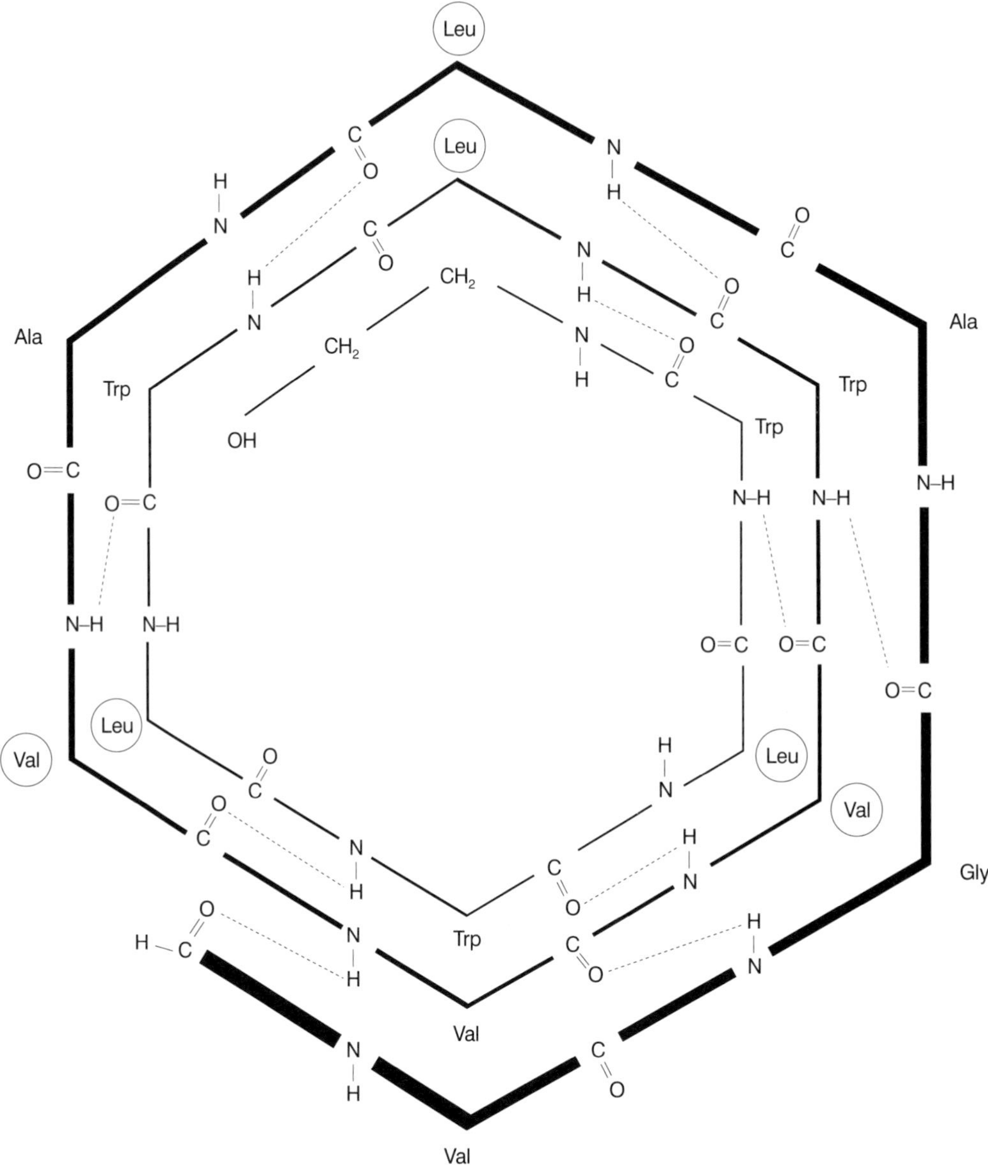

FIGURE 3.7 Gramicidin A. One possible helical structure having 6.3 residues per turn. Bonds drawn inwards are directed down the helix; those drawn outwards are directed up. D-Amino acid residues are circled.

related to gramicidin A, and lasalocid have been shown to have a surprising degree of selective toxicity for the malarial protozoal parasite *Plasmodium falciparum*. The sensitivity of the organism was studied during its intra-erythrocytic stage. The normal permeability characteristics of the red cell membrane are drastically altered by the intracellular presence of the parasite: permeability to Na^+ and Ca^{2+} ions is increased whereas permeability to K^+ is decreased. The selective toxicity of the ionophores may be due to preferential partitioning into the cell membranes of infected red cells leading to abnormal ion fluxes into the intra-erythrocytic environment of the malarial protozoan. It remains to be seen whether the interesting *in vitro* activities of the ionophores can be translated into safe and effective treatments for malarial infection.

3.4 Antifungal agents that interfere with the function and biosynthesis of membrane sterols

3.4.1 Polyene antibiotics

The polyenes constitute a large group with varied molecular structures which interact with membranes in an especially interesting way. There are about 200 polyenes, all produced by *Streptomyces* spp. Of these only a few are sufficiently non-toxic to use clinically and only one is used to treat systemic infections in man – amphotericin B (Figure 3.8). Polyenes are active against yeasts and fungi. Infections by these organisms are dangerous and amphotericin B can halt infections which might otherwise be fatal. Nystatin (Figure 3.8) is useful as a topical agent to treat localized candidal infections on mucosal surfaces. The polyenes are not absorbed from the gastrointestinal tract but are sometimes given by mouth to combat fungal growth in the gut. This most commonly results from the use of broad-spectrum antibacterials which deplete the normal bacterial flora of the gut and allow yeasts and fungi to multiply and become an opportunistic infection.

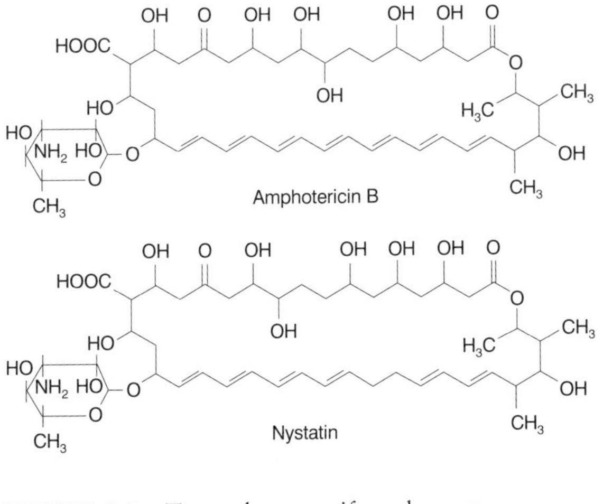

FIGURE 3.8 Two polyene antifungal agents.

The primary site of interaction is the fungal sterol, ergosterol. The sterol composition of the fungal membrane is important in determining the sensitivity to the polyene antibiotics and it is the difference in their relative affinities for ergosterol (the major sterol of fungal membranes) and the predominant cholesterol of mammalian membranes that allows polyenes to be used clinically at all. However, the similarity between cholesterol and ergosterol means that safety margins are low with polyenes. The compounds are inherently toxic to mammals, and the side-effects of amphotericin B, such as kidney damage, limit its clinical use.

Bacteria are not affected by amphotericin B or other polyenes since their membranes do not contain sterols. The mycoplasma *Acholeplasma laidlawii* does not need sterols in its membrane, but can incorporate either cholesterol or ergosterol when they are added to the growth medium. Ergosterol-containing organisms are sensitive to amphotericin B whereas cells grown without sterols are not affected. Addition of ergosterol or digitonin to yeast cell cultures prevents amphotericin from being toxic; this results from the complexing of amphotericin by these agents. Mutant yeasts which have a block at some stage of ergosterol synthesis are resistant to amphotericin because there is no longer a target for the polyene in the cell. Fortunately this type of resistance is not clinically important.

Polyenes bind specifically to sterols having a 3β-hydroxyl group and a relatively long side chain. The larger polyenes, nystatin and amphotericin B, show a preferred interaction with 3β-hydroxysterols having a conjugated diene in the fused ring nucleus. This accounts for their partial selectivity for the ergosterol-containing membranes of fungi. Nevertheless, irreversible tissue damage is always a risk during systemic therapy with amphotericin B. This toxicity can be reduced to some extent by administering the drug as a complex with cholesterol and phosphatidylglycerol.

The general action of polyenes is to increase the permeability of fungal membranes but the specific effects of individual polyenes show considerable differences. Filipin, for example, causes gross disruption of membranes with release of both solutes of low molecular weight and small proteins, whereas *N*-succinylperimycin more specifically induces the release of intracellular K^+ ions. The clinically useful drugs amphotericin B and nystatin collapse the proton motive force across the membrane by permeabilizing it to protons. This may be associated with the interaction of the polyenes with the micro-environment of the ATPase of the cytoplasmic membrane which is involved in proton transport across the membrane. The loss of K^+ from the cell may be secondary to the destruction of the proton motive force since the latter is essential for the maintenance of a high intracellular K^+ concentration. Living cells cannot survive a catastrophic loss of intracellular potassium, so the interaction of polyenes with cell membranes soon results in cell death. The ion-permeability enhancing effects of the polyenes are probably caused primarily by the drug molecules creating pores in the membranes. Molecular models of amphotericin B and nystatin show a rod-like structure held rigid by the all-*trans* extended conjugated system which is equal in length to an ergosterol molecule. The cross-section of the polyene structure is roughly rectangular. One surface of the rod is hydrophobic and the opposite surface, studded with axial hydroxyl groups, is hydrophilic. At one end of the rod the mycosamine sugar group and the carboxyl group form a zwitterionic assembly with strongly polar properties.

An attempt was recently made to construct a computer-based, 'virtual' model of a possible pore or channel involving amphotericin B and cholesterol (the investigators chose this sterol because of their interest in the toxicology of amphotericin B). The model consisted of eight amphotericin B and eight cholesterol molecules and the computer also simulated the environment of the membrane with layers of phospholipid surrounding the channel. The length of channel was less than the thickness of typical cytoplasmic membranes and it is possible that in reality two such channels end-to-end may actually bridge the membrane. It is also thought possible that the lipids around the channel may accommodate themselves so that the lipid bilayer is 'pinched in' somewhat, with the lipid bilayer now approximating to the channel length. The stability of the computer model of the channel is largely dependent upon hydrogen bonding between (a) the hydroxyl groups of neighbouring amphotericin B molecules and (b) between the amino and carboxyl groups of adjacent drug molecules. The hydroxyl groups line the internal surface of the channel and provide the necessary hydrophilic environment for the passage of K^+ and other water-soluble ionized species. Surprisingly, the computer simulation did not reveal any structurally specific interactions between the sterol and amphotericin B, although the cholesterol molecules are essential to the formation of the channel. Further details of this fascinating, though speculative, approach to the puzzle of the amphotericin B membrane channels are available in a research paper included under 'Further reading'. Considerably more research will be needed before the precise details of these channels are revealed. Furthermore, the formation of pores does not satisfactorily explain the gross permeability changes brought about by polyenes such as filipin. In this case it seems likely that the antibiotic causes a more general disruption of membrane function.

3.4.2 Inhibition of ergosterol biosynthesis

Whereas the polyenes disrupt membrane function through direct interaction with membrane ergosterol, there are several groups of antifungal compounds that act by inhibiting the biosynthesis of this sterol. An outline of the biosynthetic pathway from squalene to ergosterol is shown in Figure 3.9 and indicates the points of inhibition of several types of antifungal agents useful either in medicine or in agriculture.

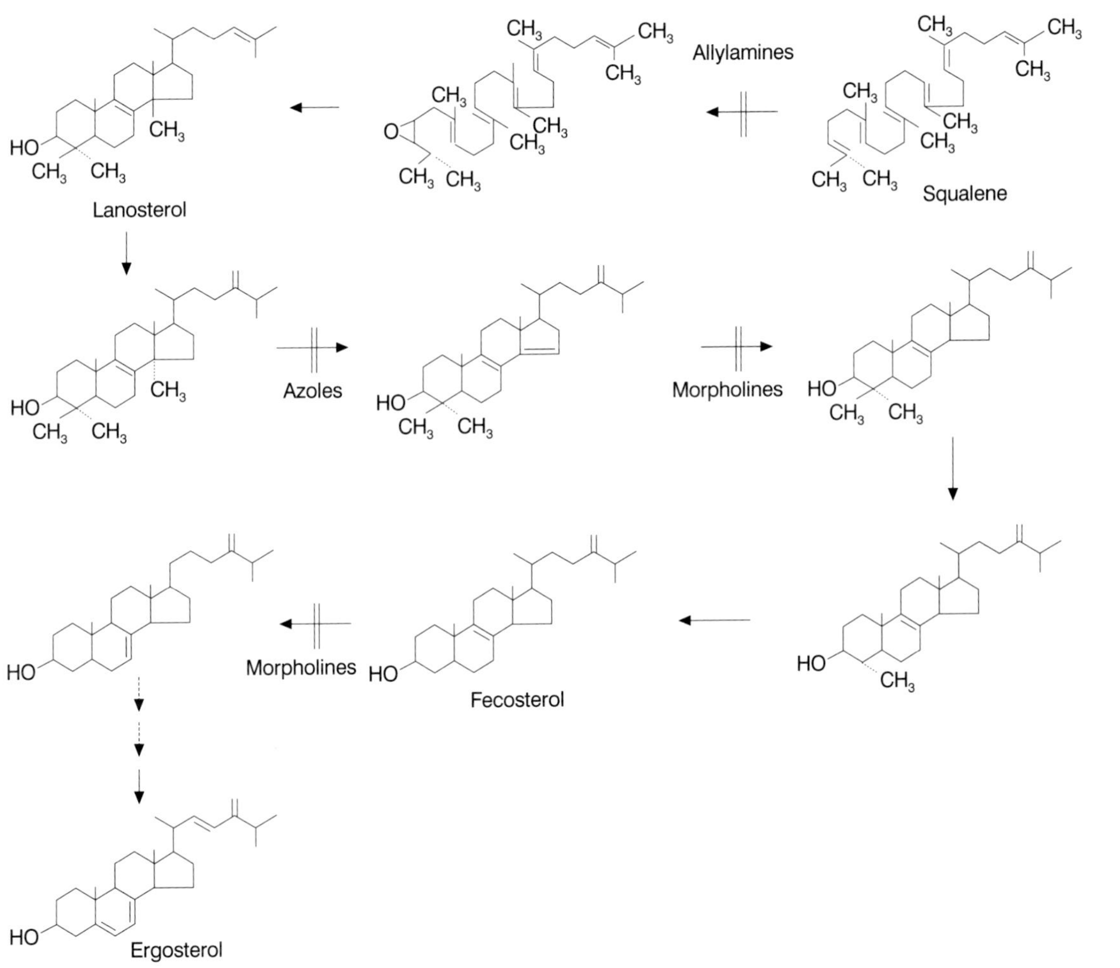

FIGURE 3.9 An outline of the biosynthesis of ergosterol from squalene, showing the points of inhibition of several types of antifungal agent.

Azoles

The azoles are among the most important compounds currently in use against fungal infections. They are subdivided into imidazoles or triazoles, according to whether they have two or three nitrogen atoms in their five-membered azole ring (Figure 3.10). Unlike the polyenes, the azoles do not kill fungi but act rather as fungistatic agents. In their favour is that they are relatively non-toxic, with the exception of some of the older compounds, such as ketoconazole which has caused fatal liver damage on rare occasions.

The antifungal action of the azoles depends on inhibition of the C-14 demethylation reaction in the biosynthesis of ergosterol. The enzyme involved, 14α-sterol demethylase, is a P_{450} cytochrome protein which the azoles inhibit by forming a stoichiometric complex with the iron of the haem component of the enzyme. The

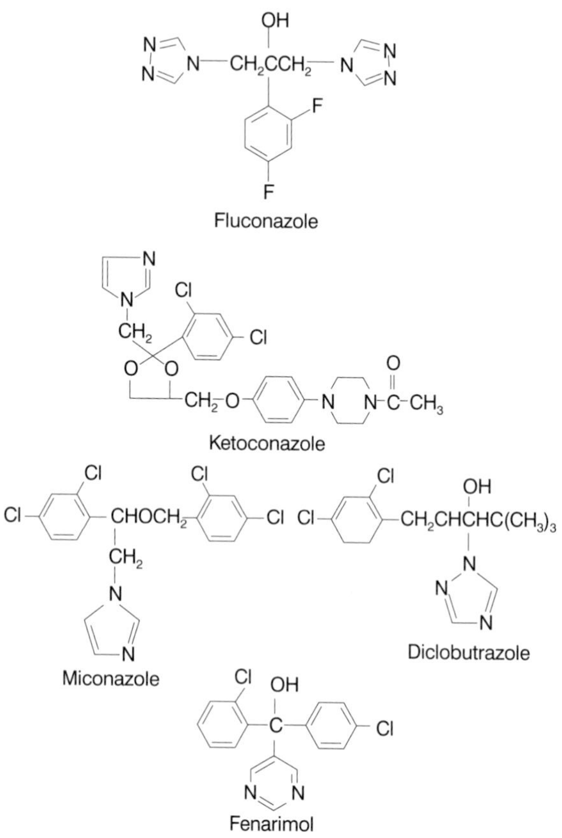

FIGURE 3.10 Azole antifungal agents used in medicine and in agriculture. Note the presence of either two or three nitrogen atoms in the heterocyclic rings. Although fenarimol has a similar biochemical action to the azoles it is, in fact, a pyrimidine derivative.

formation of the complex is detected by the red shift of the Soret band of haem from 417 nm to 447 nm. The haem interacts with the lone pair of electrons on one of the ring nitrogens and the complex is further stabilized by interactions between hydrophobic moieties in the azole ligand and the enzyme. The azole-mediated inhibition of the C-14 demethylase is non-competitive for the sterol substrate and leads to a greater net reduction in flow through the metabolic pathway than competitive inhibition. The result is an accumulation of methylated sterols in the cell and a reduction in the ergosterol content. Methylated sterols are more bulky than ergosterol and do not fit easily into a normal membrane structure. This interference in the membrane structure is thought to have adverse effects on membrane-bound enzymes such as those concerned with chitin synthesis and nutrient uptake, either directly on their activity or on their control. The depletion of ergosterol may also result in interference with its hormone-like actions on cell growth.

Typical examples of azole antifungals include the topically active agent miconazole (Figure 3.10), which is effective against thrush and dermatophyte infections, and ketoconazole (Figure 3.10), which is orally active and has been used to treat a wide range of fungal infections, particularly deep-seated, potentially life-threatening mycoses. Ketoconazole has now largely been replaced by fluconazole (Figure 3.10), which is used extensively to treat the candidal infections common in immunosuppressed patients. Because the azoles have some affinity for mammalian P_{450}-dependent enzymes, including those involved with steroid hormone synthesis, problems can arise during therapy due to depletion of testosterone and glucocorticoids. The azole diclobutrazole and the pyrimidine derivative fenarimol (Figure 3.10), which also inhibits ergosterol biosynthesis by the same mechanism as the azoles, have been used in agriculture for treating fungal infestations of plants.

Morpholines

These compounds are too toxic for systemic use in medicine because of major interference with host sterol biosynthesis; they therefore find their application as agricultural fungicides. The morpholines inhibit two stages of the ergosterol biosynthetic pathway. The first target is the enzyme that catalyses the reduction of the double bond at the 14–15 position formed after the removal of the C-14 methyl group. The second target is the isomerization of the double bond between C-8 and C-9 of fecosterol to a position between C-8 and C-7 (Δ^7–Δ^8 isomerase). The balance between these two inhibitory activities varies from fungus to fungus

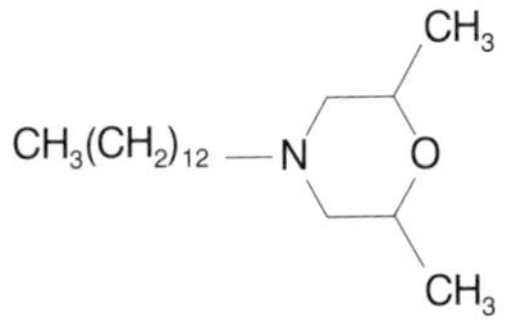

FIGURE 3.11 Tridemorph, a morpholine antifungal. Morpholines are too toxic for use in medicine and are used only in agriculture.

and probably reflects subtle differences in the enzymes involved. Tridemorph (Figure 3.11) inhibits *Ustilago maydis* mainly at the C-14 reduction step whereas *Botrytis cinerea* is inhibited mainly at the C-8-C7 isomerization. However, in general the more important target is likely to be the Δ^{14} reduction since this enzyme is essential for fungal viability whereas the Δ^7-Δ^8 isomerase is not.

Allylamines

This is another group of compounds (Figure 3.12) that affect ergosterol biosynthesis but at an earlier stage. Naftifine, a topical agent, and terbinafine, an orally active agent, are used to treat dermatophyte infections in humans and domesticated animals. These agents, along with the topical antidermatophyte compound, tolnaftate, share the same mode of action by inhibiting squalene epoxidase. Squalene accumulates in the cell with a reduction in cellular ergosterol content. Growth inhibition could result from either action. These compounds are much less active against yeasts such as *Candida albicans.*

Antifungals in current clinical use which inhibit ergosterol biosynthesis show marked selectivity for fungal systems. The azole antifungals are several hundred times more potent against lanosterol demethylation in fungi than against the corresponding reaction in mammals. Similarly, naftifine is several hundred times more potent against fungal squalene epoxidase. This species selectivity is critical in making the ergosterol biosynthesis inhibitors such good antifungal agents.

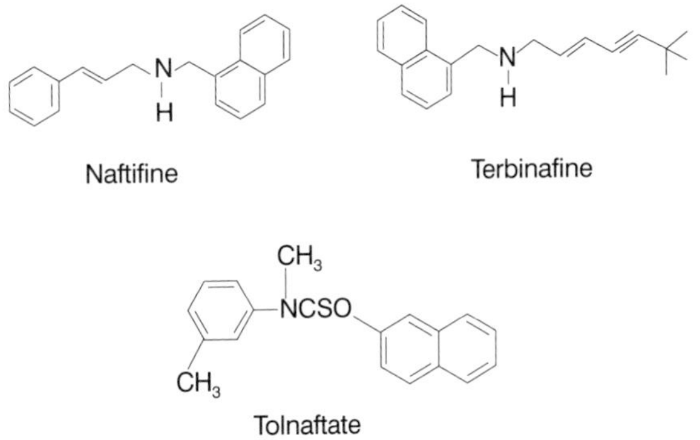

Naftifine

Terbinafine

Tolnaftate

FIGURE 3.12 Allylamine antifungal agents used to treat dermatophytic infections.

Further reading

Anderson, O. S. (1984). Gramicidin channels. *Ann. Rev. Physiol.* **46**, 531.

Baginski M. *et al.* (1997). Molecular properties of amphotericin B membrane channel: a molecular dynamics simulation. *Molec. Pharmacol.* **52**, 560.

Bernheimer, A. W. and Rudy, B. (1986). Interactions between membranes and cytolytic peptides. *Biochim. Biophys. Acta* **864**,123.

Dobler, M. (1981). *Ionophores and Their Structures,* John Wiley and Sons.

Georgopapadakou, N. H. and Walsh, T. J. (1996). Antifungal targets: chemotherapeutic targets and immunologic strategies. *Antimicrob. Agents Chemother.* **40**, 279.

Hugo, W. B. (1992). Mode of action of non-antibiotic antibacterial agents. In *Pharmaceutical Microbiology,* 5th edn, (eds. W. B. Hugo and A. D. Russell), Blackwell, Oxford, p. 288.

Ingram, L. O. and Buttke, T. M. (1984). Effects of alcohols on micro-organisms. *Adv. Microb. Physiol.* **25**, 254.

Russell, A. D. (1986). Chlorhexidine, antibacterial action and bacterial resistance. *Infection* **14**, 212.

Scott, F. M. and Coleman, S. P. (1992). Chemical disinfectants, antiseptics and preservatives. In *Pharmaceutical Microbiology,* 5th edn, (eds W. B. Hugo and A. D. Russell), Blackwell, Oxford, p. 231.

Van den Bossche H., Marichal, P. and Odds, F. C. (1994). Molecular mechanisms of drug resistance in fungi. *Trends Microbiol.* **2**, 293.

Chapter four

Inhibitors of nucleic acid synthesis

Many antimicrobial substances, both synthetic chemicals and natural products, inhibit the biosynthesis of nucleic acids. However, few of these inhibitors are clinically useful as antimicrobial drugs because most do not distinguish between nucleic acid synthesis in the infecting micro-organism and in the host. Many inhibitors of nucleic acid synthesis are therefore too toxic to the host for safe use as antimicrobial agents. However, there are important exceptions which are described in this chapter.

The synthesis of DNA and the various types of RNA is an essential function of dividing and growing cells. Inhibition of DNA synthesis rapidly results in inhibition of cell division. The biosynthesis, recombination and intercellular exchange of extrachromosomal elements of DNA in bacteria are also critical in maintaining the flexible responses of bacteria to changes in the environment (Chapter 8). Inhibition of RNA synthesis is followed by cessation of protein synthesis. The time elapsing between the inhibition of RNA synthesis and the resulting failure of protein biosynthesis can be used to indicate the stability of messenger RNA in intact cells.

Substances that interfere with nucleic acid biosynthesis fall into several categories. The first group includes compounds that interfere with the synthesis and metabolism of the 'building blocks' of nucleic acids, i.e. the purine and pyrimidine nucleotides. Interruption of the supply of any of the nucleoside triphosphates required for nucleic acid synthesis blocks further macromolecular synthesis when the normal nucleotide precursor pool is exhausted. Structural analogues of purines and pyrimidines and their respective nucleosides disrupt the supply of correct nucleotides for nucleic acid synthesis and may also inhibit nucleic acid polymerization directly following conversion to the corresponding triphosphates, either by inhibiting polymerase activity or by causing premature chain termination. Few such compounds are useful as antibacterial drugs because of their lack of specificity, but several purine and pyrimidine analogues have achieved success as antiviral agents, and a pyrimidine analogue finds application as an antifungal drug. Compounds that interfere with the supply of folic acid also inhibit nucleotide biosynthesis. Interruption of the supply of tetrahydrofolate soon brings nucleotide and nucleic acid synthesis to a halt and inhibitors of dihydrofolate reductase are useful in antibacterial and antimalarial therapy.

Although DNA-dependent RNA polymerases are common to both prokaryotic and eukaryotic cells, several naturally occurring and semi-synthetic antibiotics specifically inhibit the bacterial forms of these enzymes. Another group of inhibitors blocks nucleic acid synthesis by binding to the DNA template. This type of interaction can prevent both DNA replication and transcription into RNA, but this is often too non-specific to permit broad therapeutic application.

Finally, several series of compounds, known as topoisomerase inhibitors, block topological changes in bacterial DNA that are essential for the organization and functioning of DNA in cells. These compounds include some of the most valuable antibacterial drugs in current use.

4.1 Compounds affecting the biosynthesis and utilization of nucleotide precursors

4.1.1 The sulphonamide antibacterials

The sulphonamides were the first successful antibacterial drugs. The original observation was made with the dyestuff Prontosil rubrum, which is metabolized in the liver to the active drug sulphanilamide. A more effective derivative of sulphanilamide was sulphapyridine, which was in turn superseded by compounds with less toxic side-effects. Several of these early compounds are still in use and their structures are shown in Figure 4.1.

Many other sulphonamide antibacterials have been developed since. Most of these are probably no more intrinsically antibacterial than the earlier compounds, although some are much more persistent in the body and can therefore be dosed less frequently. The sulphonamides act against a wide range of bacteria, but their main success immediately following their discovery was in the treatment of streptococcal infections and pneumococcal pneumonia. Gradually the sulphonamides were displaced by naturally occurring antibiotics and their derivatives, largely because of the greater antibacterial potency of antibiotics. However, sulphonamides have retained a place in the treatment of certain infections, especially in combination with inhibitors of dihydrofolate reductase. The structural requirements for antibacterial activity in the sulphonamide series are relatively simple. Starting from sulphanilamide, the modifications have generally been variations in substitution on the nitrogen of the sulphonamide group. Substitution on the aromatic amino group causes loss of activity.

Among the many sulphonamides synthesized is dapsone (Figure 4.1) which, although it has no useful action against common infections, has an excellent effect on leprosy and is still a mainstay in the treatment of this disease, in combination with

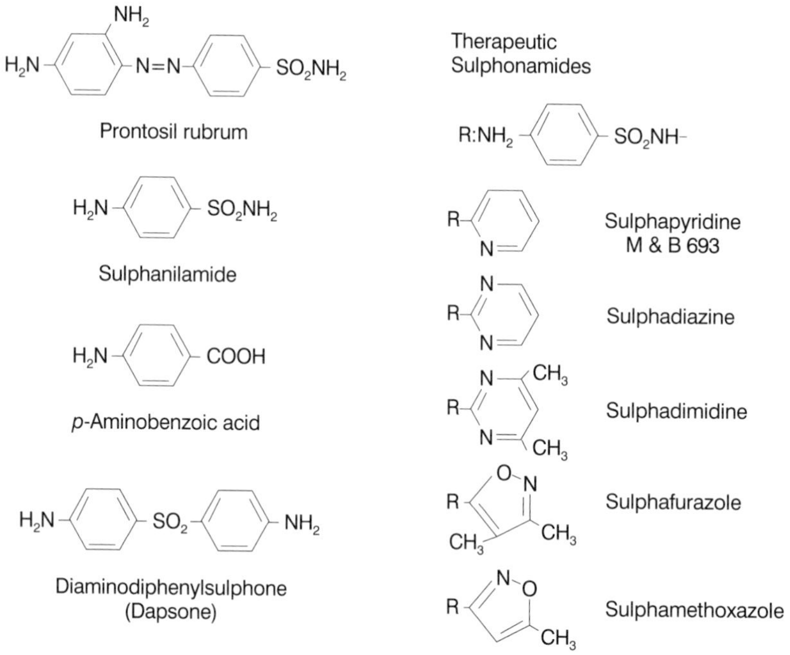

Prontosil rubrum

Sulphanilamide

p-Aminobenzoic acid

Diaminodiphenylsulphone (Dapsone)

Therapeutic Sulphonamides

R:NH_2–C$_6$H$_4$–SO_2NH–

Sulphapyridine M & B 693

Sulphadiazine

Sulphadimidine

Sulphafurazole

Sulphamethoxazole

FIGURE 4.1 Examples of sulphonamide antibacterial drugs and *p*-aminobenzoic acid.

other drugs such as rifampicin to minimize the risk of development of resistant mycobacteria. Dapsone is thought to act by the same biochemical mechanism as the sulphonamides, but the reason for its specificity in leprosy is not known.

A few years after the discovery of the antibacterial activity of the sulphonamides it was shown that some bacteria have a nutritional requirement for *p*-aminobenzoic acid, which is involved in the biosynthesis of folic acid (Figure 4.2). The structural resemblance between *p*-aminobenzoic acid and the sulphonamides underlies the ability of these drugs to antagonize the stimulatory effect of *p*-aminobenzoic acid on the growth of bacterial cells. Later, the structure of folic acid was found to contain a *p*-aminobenzyl group and its biosynthesis was shown to be inhibited by the sulphonamides. The biosynthesis proceeds via the dihydropteridine pyrophosphate derivative shown in Figure 4.2, which then reacts with *p*-aminobenzoic acid with loss of the pyrophosphate group to give dihydropteroic acid. The sulphonamides inhibit the enzyme dihydropteroate synthase which catalyses this latter reaction in an apparently competitive manner.

The sulphonamides were originally believed to compete with *p*-aminobenzoate simply by binding more tightly than the substrate to the active site of the enzyme dihydropteroate synthase but without taking part in the enzymic action. It was then found that they can also act as alternative substrates, giving rise to reaction products that are analogues of dihydropteroate. However, these products probably do not play a major role in antibacterial action since they only inhibit the downstream enzymes at concentrations higher than those achievable in the cell.

The striking success of the sulphonamides as antibacterials, coupled with the early knowledge of their point of action, led to an extraordinary flurry of chemical research. Every conceivable bacterial growth factor became the model for the synthesis of analogues in the hope of repeating the success of the sulphonamides as antibacterial agents. Unfortunately this tremendous effort was largely fruitless because the apparently simple model provided by the antagonism of *p*-aminobenzoic acid by sulphanilamide was not easily repeated. The sulphonamides owe their effect to a fortunate set of circumstances: *p*-aminobenzoate is not a metabolic intermediate in animal cells, which acquire their folic acid from the diet, and the inhibition of bacterial growth is not reversed by folic acid because of its poor diffusion into the cells. In contrast, the sulphonamides, like *p*-aminobenzoic acid, enter bacterial cells freely. Many biosynthetic intermediates carry phosphoric acid groups which tend to prevent their diffusion into bacteria, and potential inhibitors based on analogous structures share the same difficulty of access. This problem was not readily appreciated during the early days of antibacterial drug research.

4.1.2 Inhibitors of dihydrofolate reductase

When the structure of folic acid became known and its relationship to *p*-aminobenzoic acid and the sulphonamides was accepted, a search was made for antagonists among structural analogues of folic acid itself. These were found, but not surprisingly they were highly toxic because folic acid derivatives, in contrast with *p*-aminobenzoic acid, play an important part in the metabolism of animal cells. The toxicity of some of these compounds towards animal cells is actually much greater than towards bacteria since bacterial membranes are almost completely impermeable to them. The cytotoxic action of the antifolic compound methotrexate (Figure 4.3), has found a practical application in the treatment of certain malignancies, rheumatoid arthritis and psoriasis.

Although the direct analogues of folic acid were of no value as antibacterial agents, other compounds more distantly related to folic acid have considerable importance. The potential of this type of compound was first realized in two drugs developed as antimalarials, pyrimethamine and proguanil (Figure 4.3). The latter compound is a prodrug that is metabolized in the liver to the active agent, cycloguanil.

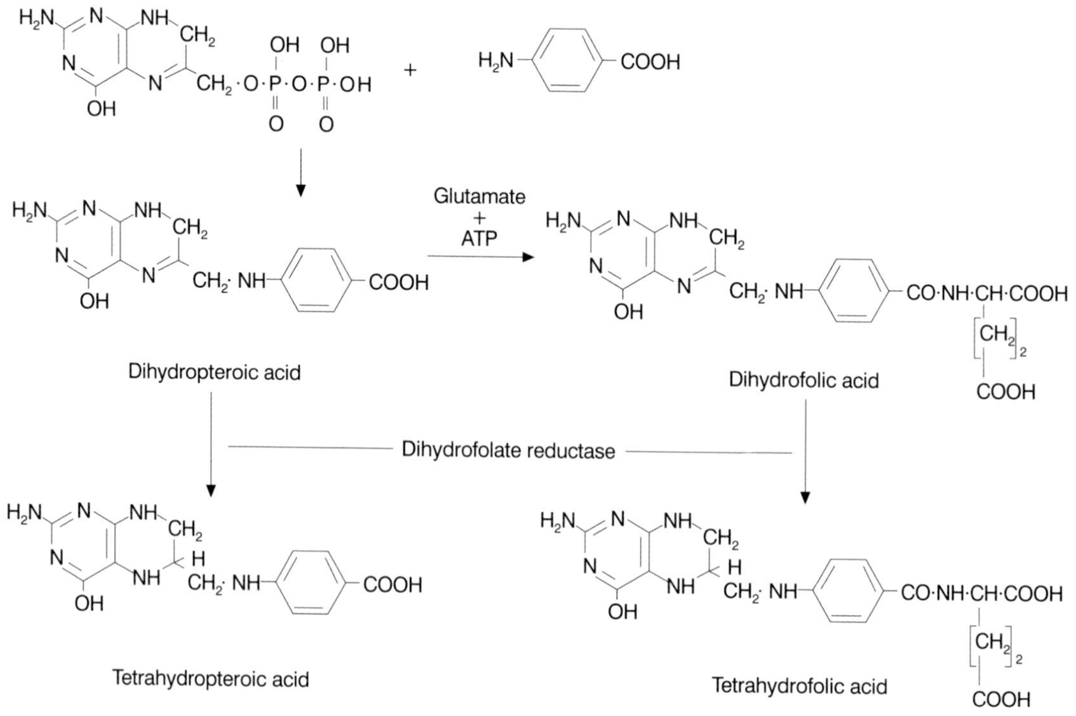

Glutamate + ATP

Dihydropteroic acid

Dihydrofolic acid

Dihydrofolate reductase

Tetrahydropteroic acid

Tetrahydrofolic acid

FIGURE 4.2 The final stages of folic acid biosynthesis. The first reaction in the sequence is catalysed by dihydropteroate synthase which is competitively inhibited by sulphonamides.

The exact point of attack of these so-called antifolic compounds became apparent when the details of folic acid biosynthesis were fully worked out. The step leading to the production of dihydropteroic acid has already been discussed. At this point glutamic acid is added to give dihydrofolic acid which must be reduced to the tetrahydro state by the enzyme dihydrofolate reductase (Figure 4.2) before it can participate as a cofactor for one-carbon transfer reactions. Cytotoxic analogues of folic acid, such as methotrexate and the antimalarial drugs mentioned above, inhibit dihydrofolate reductase. Although most living cells contain dihydrofolate reductase, the enzyme evidently differs in structural details amongst major groups of organisms, and a useful degree of species specificity in the action of inhibitors is possible. For example, pyrimethamine is poorly active against the mammalian and bacterial enzymes but has an exceptionally strong affinity for the enzyme from the malarial parasite, which accounts for its specific antimalarial action. The antimalarial metabolite of proguanil presumably has analogous specificity for the protozoal dihydrofolate reductase. A highly selective compound against the bacterial dihydrofolate reductase is the pyrimidine derivative trimethoprim (Figure 4.3). Reduction of the activity of bacterial dihydrofolate reductase by 50% requires a trimethoprim concentration of 0.01 μM, whereas the same inhibition of the human enzyme requires 300 μM. Trimethoprim is effective on its own as an antibacterial drug but is more generally used as a combination (co-trimoxazole) with the sulphonamide derivative sulphamethoxazole (Figure 4.1). The combination is claimed to have a wider field of antibacterial activity than either compound alone and is prescribed as a broad-spectrum alternative to ampicillin. Since both the sulphonamide and trimethoprim block the folic acid biosynthetic pathway, but at different points,

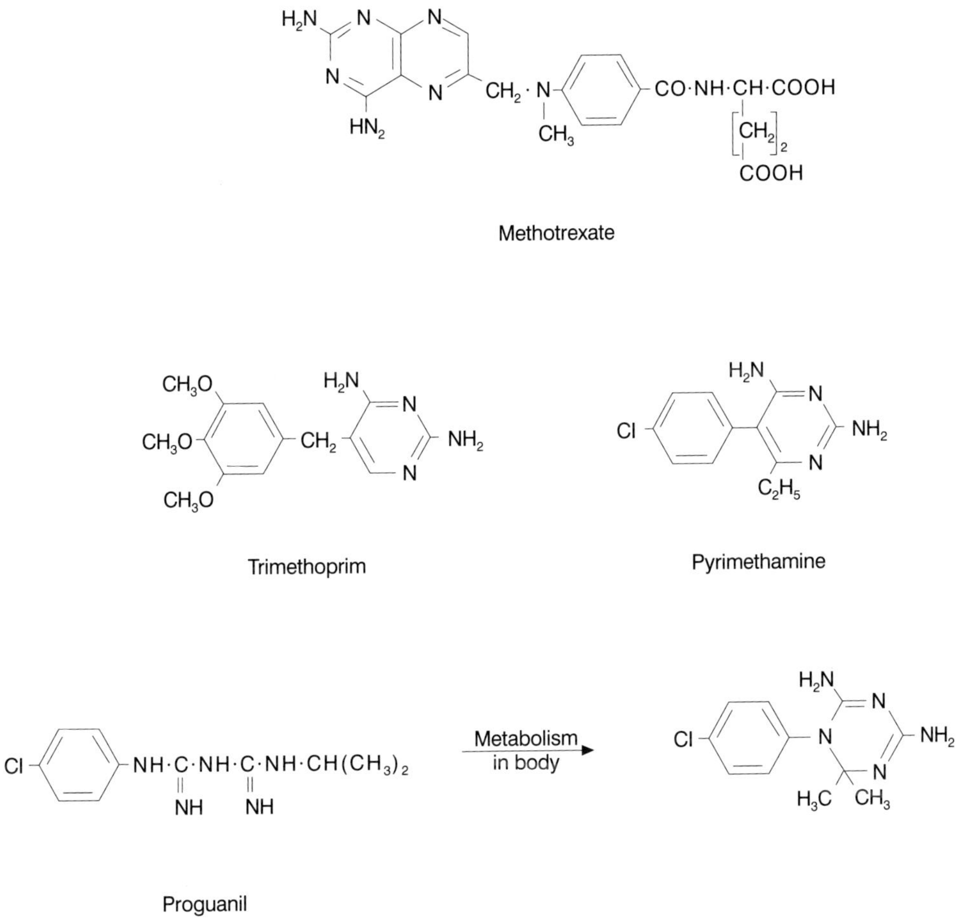

Methotrexate

Trimethoprim

Pyrimethamine

Metabolism in body

Proguanil

FIGURE 4.3 Drugs that inhibit dihydrofolate reductase. In the case of the antimalarial proguanil, its metabolite, cycloguanil, is the active inhibitor.

the twofold blockage is especially effective in depriving bacteria of tetrahydrofolate.

The reduction in tetrahydrofolate levels in bacteria caused by sulphonamides and the dihydrofolate reductase inhibitors has widespread effects on the cell. Tetrahydrofolate is required as a one-carbon-unit donor in the biosynthesis of methionine, glycine and the formyl group of fMet-tRNA, thus tetrahydrofolate deprivation results in reduced protein synthesis. The major effects, however, are on the biosynthesis of purines and pyrimidines, which involve one-carbon transfer reactions at several stages. The synthesis of thymine is particularly sensitive to inhibitors of dihydrofolate reductase because of the requirement for tetrahydrofolate in the transformation of dUMP to dTMP (Figure 4.4). When cultures of bacteria are grown in media containing amino acids and inosine, antagonists of folic acid synthesis cause the phenomenon known as 'thymineless death', which can be prevented by the addition of excess thymine or thymidine.

4.1.3 Nucleoside analogues

5-Fluorocytosine (flucytosine, 5FC)

This pyrimidine analogue (Figure 4.5) was originally synthesized as an anticancer agent but is now used mainly to treat certain serious fungal infections,

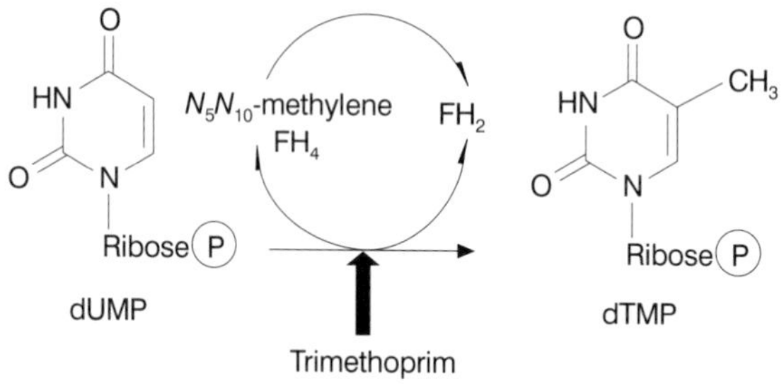

FIGURE 4.4 A major consequence of the inhibition of dihydrofolate reductase is the suppression of thymine biosynthesis. Bacteria cultured with trimethoprim undergo 'thymineless death' when not supplemented with thymine or thymidine.

including cryptococcosis, candidiasis and chromomycosis. Except in the latter infection, 5FC is usually administered in combination with amphotericin B (Figure 3.8). 5FC is not itself active and must be metabolized to compounds that are the effective inhibitors. The uptake of 5FC into fungal cells is facilitated by the transporter protein, cytosine permease. Subsequently, the compound is converted to 5-fluorouracil by cytosine deaminase, fortunately an enzyme that is absent from human cells since 5-fluorouracil is highly toxic to all dividing mammalian cells. 5-Fluorouracil enters a complex network of fungal nucleotide metabolism. An important end-product is 5-fluorodeoxyuridine-5′-monophosphate, which inhibits thymidylate synthase and therefore DNA synthesis. The other major route of metabolism that contributes to the antifungal activity of 5FC is via the conversion to 5-fluorouridine-5′-monophosphate, catalysed by UMP phosphoribosyl transferase, leading to the incorporation of 5-fluorouridine-5′-triphosphate into RNA. The antifungal action of 5FC therefore results from a combination of the inhibition of DNA synthesis and the generation of aberrant RNA transcripts.

During therapy with 5FC careful monitoring of blood levels is important to ensure that concentrations toxic to the kidneys and bone marrow are not achieved, but that levels are nevertheless high enough to minimize the risk of the emergence of 5FC-resistant mutants.

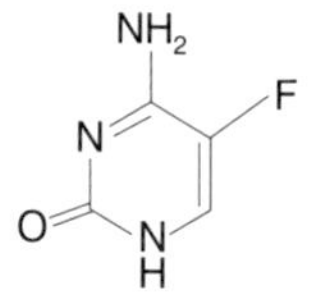

FIGURE 4.5 The antifungal drug, 5-fluorocytosine.

Ribavirin

In the past this nucleoside analogue (1-β-D-ribofuranosyl-1,2,4-triazole-3-carboxamide; Figure 4.6) has been used mainly to treat serious lung infections in young children, caused by the respiratory syncytial virus. In combination with interferon-α, ribavirin is now finding an additional application in the treatment of hepatitis C. The basis of the antiviral action of ribavirin is somewhat controversial, mainly because of the many potential points of attack in nucleotide metabolism. The favoured view is that the monophosphate derivative competitively inhibits the host enzyme inosine-5′-monophosphate dehydrogenase. This enzyme, which is not encoded by viruses, catalyses the conversion of inosine-5′-monophosphate to xanthosine-5′-monophosphate, the rate-limiting step in the biosynthetic sequence leading to guanine nucleotides. Depletion of the cellular pool of the guanine ribo- and deoxyribonucleotides eventually halts the synthesis of viral nucleic acids. Inevitably, however, there is concomitant interference with host nucleic acid synthesis and some associated toxicity.

4.2 Inhibitors of nucleic acid biosynthesis at the polymerization stage

4.2.1 Antiviral nucleoside analogues

Several nucleoside analogues have achieved considerable success as antiviral drugs. Like 5FC, all these compounds are prodrugs that are metabolized by host or viral enzymes to the active inhibitors. The antiviral action of nucleoside analogues is probably due to a combination of interference with nucleotide metabolism and nucleic acid polymerases.

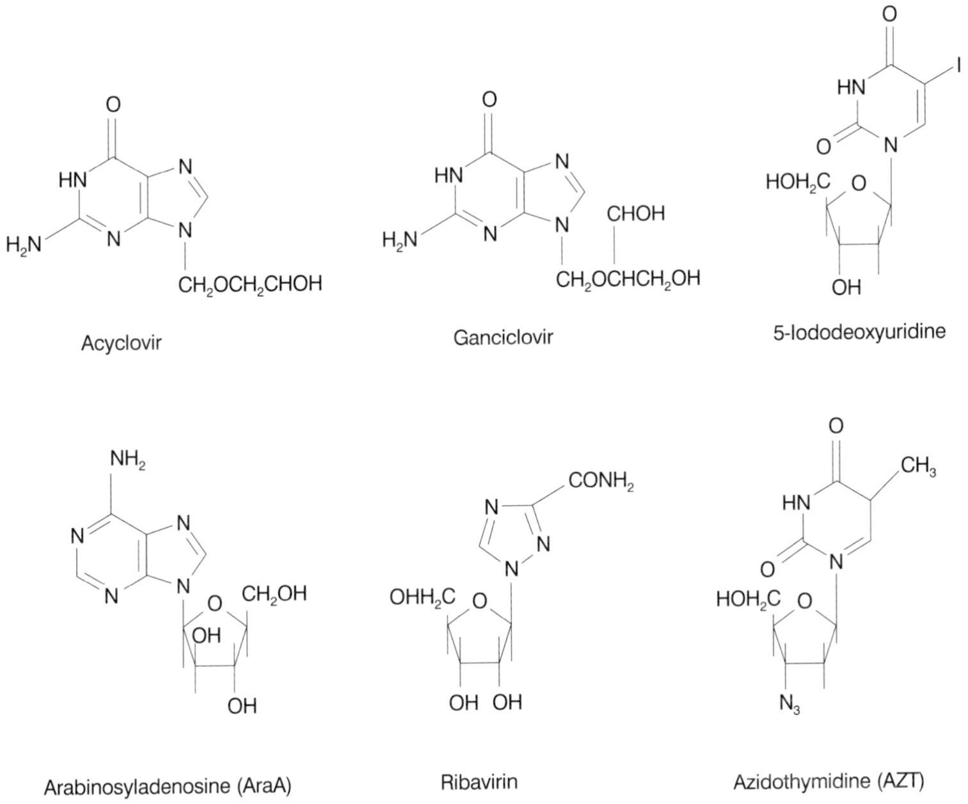

FIGURE 4.6 Examples of nucleoside analogues as antiviral drugs.

Acyclovir and ganciclovir

These structurally similar compounds (Figure 4.6) are analogues of guanosine but lacking the cyclic ribose group. Despite their similarity, the two drugs have different clinical applications. Acyclovir is used to treat herpes simplex and varicella-zoster infections. While ganciclovir is also active against herpes viruses, its clinical application is limited to the treatment of cytomegalovirus infections, which are particularly troublesome in AIDS patients.

Herpes viruses code for a virus-specific form of thymidine kinase which converts acyclovir and ganciclovir to their monophosphate derivatives. Significantly, the thymidine kinase of the host cells has a much lower substrate affinity for these compounds so that uninfected cells do not generate the phosphorylated derivatives. The drug monophosphates in virus-infected cells are then successively phosphorylated by host cell kinases to the triphosphate level. The triphosphates of both drugs are good substrates for the DNA polymerases encoded by herpes viruses but are only poorly recognized by the host polymerases. The resulting preferential incorporation of the drug triphosphates into viral DNA causes premature chain termination since the antiviral nucleotides lack a 3′-OH group on the acyclic sugar residues, thus preventing formation of the 3′-5′-phosphodiester bonds necessary for chain extension.

The action of ganciclovir against cytomegalovirus is rather different. In cytomegalovirus-infected cells the drug is again first converted to the monophosphate, though not by thymidine kinase, which cytomegalovirus lacks. Instead, it is believed that another viral kinase, encoded by the *UL97* gene, may be responsible for the first-stage phosphorylation. The subsequent stages in the metabolism of ganciclovir monophosphate and its antiviral action resemble those in herpes-infected cells.

Vidarabine

Vidarabine (9-β-D-arabinosyladenine, AraA; Figure 4.6) has a fairly broad antiviral spectrum, including activity against herpes viruses, cytomegalovirus and the Epstein–Barr virus. Its triphosphate metabolite inhibits viral DNA polymerase, viral ribonucleotide reductase and is also incorporated into viral DNA. Because all of these effects occur at concentrations of vidarabine below those needed to inhibit host DNA synthesis, the compound can be used both topically and systemically against herpes infections of the eye and brain.

5-Iododeoxyuridine

5-Iododeoxyuridine (Figure 4.6) is another nucleoside which has been used against herpes infections, especially of the cornea, but is less specifically antiviral than the compounds described above. The similar Van der Waals radii of the iodine atom (0.215 nm) and the methyl group (0.2 nm) of thymidine enable the drug to replace thymidine in DNA with considerable efficiency, since its triphosphate is readily accepted as a substrate by DNA polymerase. The incorporation of 5-iododeoxyuridine into viral DNA leads to errors of replication and transcription and the eventual termination of viral proliferation. Unfortunately the compound is also incorporated into host DNA and the resultant toxicity limits its use to topical application to the eye.

4.2.2 Nucleoside analogues active against the human immunodeficiency virus (HIV)

The search for effective treatments against the virus that causes AIDS has been, and continues to be, one of the greatest therapeutic challenges of the twentieth century. The progress of the infection and its associated pathology is insidious and irreversible damage to the immune system may occur if the disease is not diagnosed sufficiently early. Nevertheless, considerable progress has been achieved in developing drugs that limit viral proliferation and confer significant clinical benefits. Nucleoside analogues have made a notable contribution to this encouraging development. A completely different approach, based on the inhibition of an HIV-specific protease involved in virus assembly, is described in Chapter 6.

Azidothymidine (3′-azido-3′-deoxythymidine, AZT; Figure 4.6) was originally developed as a potential anticancer drug but proved to be the first effective agent against the replication of HIV in AIDS patients. The compound is first efficiently converted by the host cell thymidine kinase to the monophosphate derivative, AZT-MP. The subsequent phosphorylations to the di- and tri-phosphate derivatives are also carried out by a host cell enzyme, thymidylate kinase. AZT-MP is, however, a relatively poor substrate for this enzyme, in part because the larger size of the 3′-azido group of AZT-MP compared with the 3′-OH group of thymidine monophosphate hinders the interaction of AZT-MP with the active site of thymidylate kinase. The triphosphate of AZT, inhibits the virus-specific reverse transcriptase by competing with the endogenous thymidine triphosphate. The reverse transcriptase transcribes HIV RNA into DNA, which is subsequently integrated into the host genome. HIV reverse transcriptase has three distinct enzymic activities:

1. an RNA-dependent DNA polymerase;
2. a ribonuclease (H); and
3. a DNA-dependent DNA polymerase.

Respectively, these activities:

1. copy the plus-strand RNA of the virus to produce a minus-strand DNA;
2. remove the RNA template; and
3. synthesize the plus-strand of DNA using the minus-strand DNA as a template.

The same polymerase is believed to carry out both RNA- and DNA-dependent processes. The polymerase activity of reverse transcriptase can also incorporate AZT-MP into DNA which then blocks chain extension because the AZT residue is unable to form a 3'-5'-phosphodiester bond.

Because there is no reverse transcriptase in uninfected host cells, it might first appear that AZT would be specifically active against the virus infection. Unfortunately, the clinical use of AZT is beset with problems. First, virus replication is only reduced to about 10% of normal, largely because of the relatively inefficient conversion of AZT-MP to the antiviral triphosphate. The incomplete inhibition of viral replication facilitates the emergence of AZT-resistant mutants (see Chapter 9). Secondly, there is a major problem of bone marrow toxicity, probably due to the interference of AZT-phosphates with the pyrimidine metabolism of host cells and also with host cell DNA synthesis.

Another nucleoside analogue, di-deoxyinosine, which has essentially the same mechanism of action as AZT, is now widely used in the treatment of AIDS in combination with AZT and an inhibitor of HIV protease.

Finally, mention must be made of a different class of inhibitors of HIV reverse transcriptase which are not nucleoside analogues. One of these compounds, nevirapine (Figure 4.7), inhibits the enzyme without prior metabolism and is currently undergoing clinical trials in AIDS patients. Unfortunately, nevirapine-resistant variants of the reverse transcriptase emerge readily during treatment so that the long-term clinical use of nevirapine remains in doubt.

FIGURE 4.7 The antiviral drug nevirapine, a non-nucleoside inhibitor of the reverse transcriptase of HIV.

4.2.3 Inhibitors of DNA-dependent RNA polymerase

The transcription of RNA from DNA is, of course, common to both prokaryotic and eukaryotic organisms and involves enzymes known as DNA-dependent RNA polymerases. It was therefore somewhat surprising to discover that an important group of antibacterial antibiotics, the rifamycins isolated from *Streptomyces,* show remarkable specificity for the inhibition of bacterial DNA-dependent RNA polymerase. One of these compounds, rifampicin (Figure 4.8), is a mainstay of treatment for tuberculosis in combination with other drugs. Rifampicin is active against many Gram-positive bacteria but is only poorly effective against Gram-negatives because of limited access to the target enzyme in these organisms. Chemically the rifamycins are closely related to the streptovaricins (Figure 4.8). The two groups of antibiotics appear to have a similar mode of action but, unlike the rifamycins, the streptovaricins are not in medical use. Both groups strongly inhibit RNA synthesis in sensitive bacteria and also in cell-free extracts by binding to and inhibiting DNA-dependent RNA polymerase. The drugs neither bind to nor inhibit the corresponding mammalian enzyme.

The RNA polymerase isolated from *Escherichia coli* is a large (450 kDa) complex enzyme consisting of four kinds of subunit: α, β, β' and σ. The complete or holoenzyme has the composition ($\alpha_2\beta\beta'\sigma$) together with two tightly bound zinc atoms. The function of the σ subunit is to locate a promoter site where transcription is initiated. The σ subunit then dissociates from the rest of the enzyme leaving the core enzyme ($\alpha_2\beta\beta'$) bound to the DNA template via the β' subunit. The β subunit carries the catalytic site for the internucleotide bond formation and is the target for antibiotic inhibition. This was revealed by studies with RNA

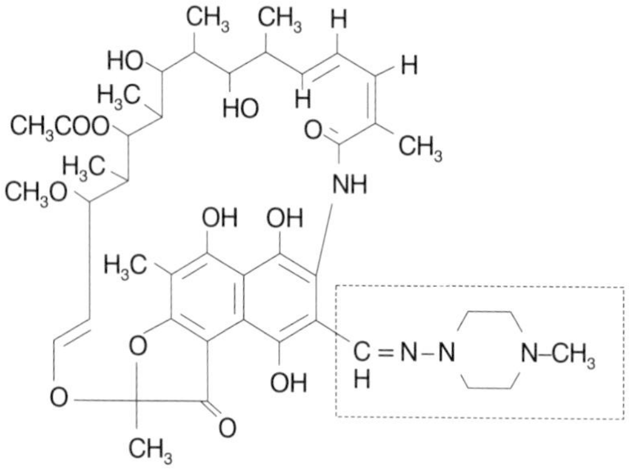

FIGURE 4.8 Two antibiotics that selectively inhibit DNA-dependent RNA synthesis in bacteria. Rifampicin is a semi-synthetic member of the rifamycin group; the synthetic side chain is enclosed by the dotted line. Streptovaricin D is related in structure to the rifamycins; jointly the rifamycins and streptovaricins are known as ansamycins.

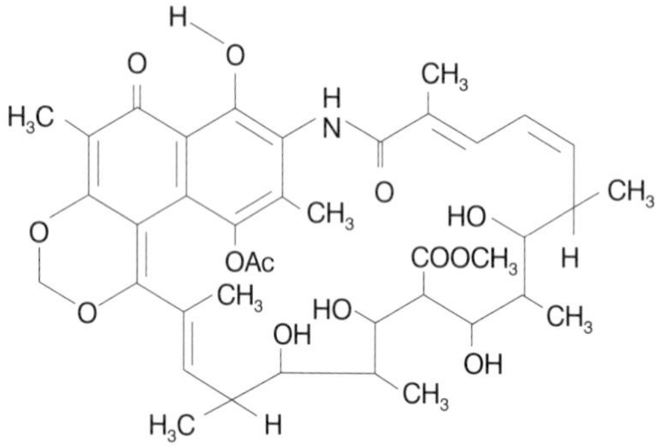

polymerase isolated from rifampicin-resistant bacteria. The molecular basis of resistance was found to be located in the β subunit of the core enzyme which fails to bind the antibiotic.

Structural analysis of the RNA polymerase from *Escherichia coli* indicates that there are two distinct substrate binding sites on the β subunit. The 'i' or initiation site, is template-independent and recognizes only purine nucleoside triphosphates. The second site, called the 'i+1' site, has no nucleotide preference. The initiation of transcription is marked by the formation of an internucleotide bond between the nucleotides bound to the i and i+1 sites. Rifampicin and structurally related antibiotics have long been known to block initiation. Fluorometric analysis of the complex between the antibiotic and enzyme suggests that rifamycin binds to the β subunit some 30 Å from the i site and approximately 20 Å from the i+1 site. The binding of the antibiotic is a two-stage process:

$$R + E \leftrightarrow RE \leftrightarrow RE^*$$

The first stage is a fast bimolecular reaction, followed by a second, slower unimolecular process involving a conformation change in the enzyme necessary for the inhibitory action of rifampicin. The overall dissociation constant for the interaction is very low, 3 nm, indicating tight but not covalent binding.

The formation of the first internucleotide bond is not inhibited by rifampicin and the major effect of the antibiotic is to block the subsequent step. Antibiotic bound to the β subunit is thought to hinder the interaction of the next incoming nucleotide with the catalytic site. However, if initiation of RNA chain synthesis progresses beyond the second or third phosphate diester bond before the addition of drug, further chain elongation is insensitive to the action of rifampicin.

Streptolydigin (Figure 4.9), like the rifamycins and streptovaricins, is a specific inhibitor of bacterial RNA polymerase. However, it inhibits chain elongation as well as the initiation process and increases the stability of purified RNA polymerase–DNA template complexes. The β subunit of the polymerase core enzyme bears the streptolydigin binding site and the increased stability of the enzyme–template–antibiotic complex delays the progress of the enzyme along the template without affecting the accuracy of the transcriptional process.

Despite the evidence obtained *in vitro* for the mode of action of streptolydigin, studies of its

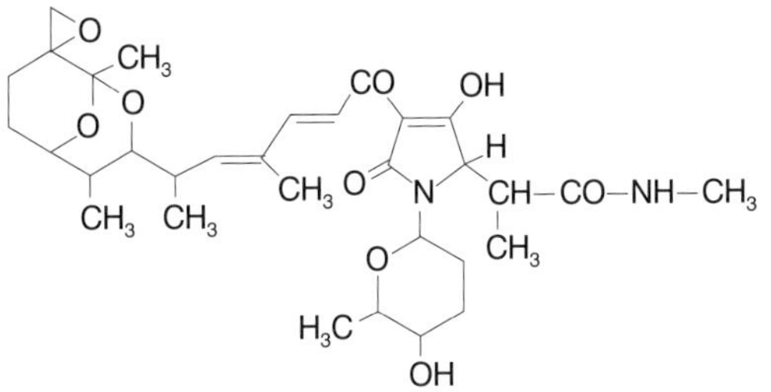

FIGURE 4.9 In addition to the ansamycins, the structurally distinct antibiotic streptolydigin also inhibits bacterial DNA-dependent RNA polymerase.

effects on intact *Escherichia coli* cells indicate that streptolydigin *in vivo* may accelerate the termination of RNA chains. The rate of elongation of RNA chains is unaffected but streptolydigin may destabilize the transcription complex *in vivo,* thus permitting premature attachment of termination factors. Only more research will resolve this apparent conflict between studies *in vitro* and *in vivo* on the mode of action of streptolydigin. Streptolydigin has not found a clinical application.

4.2.4 Inhibition of nucleic acid synthesis by interaction with DNA

As we have seen, the synthesis of nucleic acids can be interrupted by blocking the supply of essential nucleotides, by the termination of chain extension following the incorporation of some nucleotide analogues into the nucleic acid structure, and by the direct inhibition of certain polymerases. Another type of inhibition depends upon the formation of a complex between the inhibitor and DNA, thereby interfering with template function in replication and transcription. Some of the best-known examples of this group have useful anticancer activity and include actinomycin D, bleomycin and mitomycin C. These compounds are powerfully cytotoxic to mammalian and microbial cells alike, and therefore have no application in antimicrobial therapy and will not be further discussed.

However, several compounds that interact with DNA by a mechanism known as intercalation have applications as antimicrobial agents. Although it is not clear in every case that intercalation forms the primary basis for antimicrobial action, the molecular mechanism is sufficiently interesting to warrant inclusion.

Acridines, phenanthridines and choroquine

The medical history of the acridine dyes extends over some 80 years since proflavine (Figure 4.10) was used as a topical disinfectant on wounds during the First World War. Proflavine is too toxic to be used as a systemic antibacterial agent, but the related acridine, mepacrine (Figure 1.2) found wide application as an antimalarial drug before its replacement by more 'patient friendly' drugs. Chloroquine (Figure 4.10) is still an important antimalarial agent although, as we shall see in Chapter 6, its interaction with DNA is no longer considered to be the primary basis of its action against the malarial parasite. The phenanthridine ethidium (Figure 4.10) has some application as a trypanocide in veterinary medicine.

The compounds all bind to the nucleic acids of living cells and the phenomenon forms the basis of the technique known as vital staining, since the nucleic acid–dye complexes exhibit characteristic colours when examined by fluorescence microscopy. The dyes also bind readily to nucleic acids *in vitro* and the visible absorption spectra of

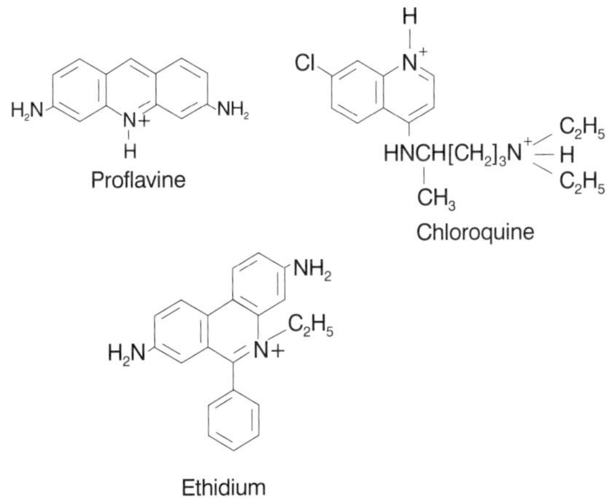

FIGURE 4.10 Three compounds that intercalate with DNA.

the ligand molecules undergo a metachromatic shift to longer wavelengths.

Two types of binding to DNA are recognized: a strong primary binding which occurs in a random manner in the molecule, and a weak secondary binding. The strong primary binding occurs only with DNA, although many other polymers bind the dyes by the secondary process. The primary binding to DNA, which is mainly responsible for the ability of the drugs to inhibit nucleic acid synthesis, depends upon the insertion or intercalation of the rigidly planar molecules between the adjacent stacked base-pairs of the double helix of DNA.

What is the evidence for this unusual type of interaction with DNA? The intercalation of a molecule into DNA is detectable by various physical changes:

1. DNA solutions show an increase in viscosity.
2. There is a decrease in the sedimentation coefficient of DNA as determined by ultracentrifugation, indicating a reduction in its buoyant density.
3. The thermal stability of DNA, i.e. the temperature at which the double helix begins to unwind, is increased.

The extent of these changes is proportional to the amount of drug intercalated into the double helix. In the case of the anticancer drug actinomycin D there is direct X-ray crystallographic evidence for intercalation into an oligo-deoxyribonucleotide.

The increase in the viscosity of DNA solutions treated with intercalating drugs is explained by the restricted degree of irregular tertiary coiling that the double helix can undergo. The DNA–drug complex is therefore both straighter and stiffer than the uncomplexed nucleic acid and these changes raise the viscosity. The reductions in sedimentation coefficient and buoyant density of DNA following intercalation result from a reduction in the mass per unit length of the nucleic acid. For example, a proflavine molecule increases the length of the DNA by about the same amount as an extra base-pair, but because proflavine has less than half the mass of the base-pair, the mass per unit length of the complexed DNA is decreased. The increased thermal stability of intercalated DNA is probably due in part to the extra energy needed to remove the bound drug from the double helix in addition to that required to separate the strands.

To permit the insertion of an intercalating molecule into DNA it is believed that a local partial unwinding of the double helix associated with the normal molecular motions within the macromolecule produces spaces between the stacked base-pairs into which the planar polycyclic molecule can move. A model advanced many years ago shows schematically how polycyclic structures may intercalate between the stacked base-pairs (Figure 4.11). The hydrogen bonding between the base-pairs remains undisturbed, although there is some distortion of the smooth coil of the sugar phosphate backbone as the intercalated molecules maintain the double helix in a partially unwound configuration. It is believed that the distortion of the double helix together with the hindrance to strand separation are major factors in blocking DNA replication and transcription.

The details of specific drug–DNA interactions depend largely on the structures of the individual

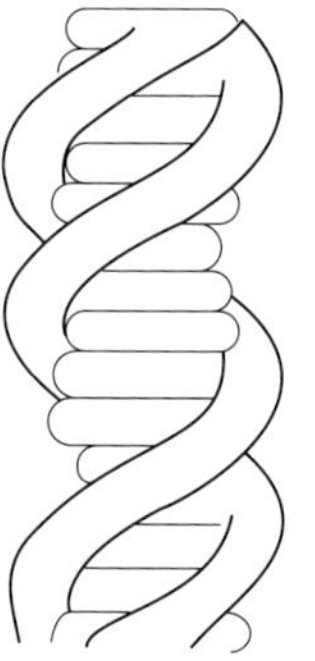
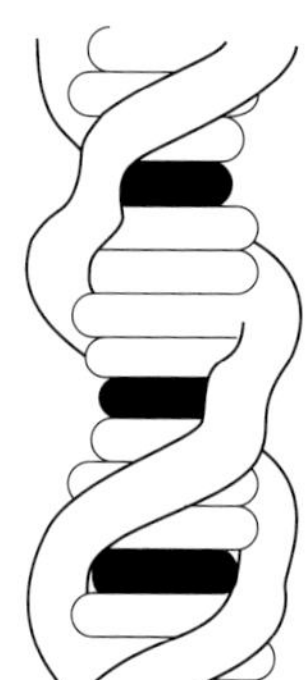

FIGURE 4.11 Diagram to represent the secondary structure of normal DNA (left) and DNA containing intercalated molecules (right). The stacked bases of the nucleic acid are separated at intervals by the intercalators (black), resulting in some distortion of the sugar–phosphate backbone of the DNA. (Reproduced with permission from L. Lerman (1964) *J. Cell Comp. Physiol.* **64**, Suppl. 1, copyright owned by Wiley-Liss, Inc., a subsidiary of John Wiley & Sons.)

drugs. In general terms, intercalation is probably stabilized by electronic interactions between the characteristic planar ring systems of the compounds and the heterocyclic bases of the DNA above and below the drug. The complexes formed by proflavine and ethidium may also be stabilized by hydrogen bonding between their amino groups and the charged oxygen atoms of the phosphate groups in the sugar–phosphate backbone. In the case of chloroquine, the projecting cationic side chain may form a salt linkage with a phosphate residue.

In certain tumour viruses and bacteriophages, in the kinetoplasts of trypanosomes and in bacterial plasmids (Chapter 8) double-stranded DNA exists as covalently closed circles. Circular DNA, covalently closed via the 3′-5′-phosphodiester bond, is characteristically supercoiled because the circular molecule is in a state of strain. The strain is relieved and the supercoils often disappear when single-stranded breaks or 'nicks' are produced by endonuclease action. Closed circular DNA has an unusual affinity for intercalating molecules which, because they partially unwind the double helix, also reduce the supercoiling of the DNA. If the unwinding proceeds beyond a certain point, as more and more drug is added, the DNA begins to adopt the supercoiled form again but in the opposite direction from that of the uncomplexed DNA. At this point the affinity of the closed circular DNA for the intercalated molecules declines until it is less than that of nicked DNA.

The diminished affinity of closed circular DNA for ethidium (Figure 4.10) at high concentrations of the drug permits a convenient separation of closed circular DNA from nicked DNA, as the sedimentation coefficient and buoyant density of DNA with a lower content of intercalated compound are significantly higher. This effect is useful in the isolation of closed circular DNA on a preparative scale.

It is also possible that the initial higher affinity of supercoiled DNA for intercalating molecules may in part account for their specificity of action against organelles and organisms that contain circular DNA. Thus the treatment of bacteria with acridines can cause the loss of plasmids from the cells. The mitochondria of certain strains of yeast are severely and irreversibly damaged by growth in the presence of ethidium, apparently due to drug-induced mutations affecting the mitochondrial DNA. The kinetoplast of trypanosomes is also seriously affected by intercalating agents, DNA synthesis in this organelle being selectively inhibited. Eventually the kinetoplast disappears altogether. As this adversely affects the life cycle of trypanosomes it is possible that the selective attack on the kinetoplast may underlie the trypanocidal activity of intercalating drugs such as ethidium.

4.2.5 Topoisomerase inhibitors

DNA cannot exist in bacterial cells as an extended double-helical molecule. The length of bacterial DNA is around 1300 μm and typically the cell into which it fits is around 1 μm in diameter. Clearly there must be a high degree of ordered quaternary structure in the DNA to accommodate it within the cells. This is achieved by negatively supercoiling the DNA, i.e. the supercoiling is left-handed in contrast with the right-handed winding of the double helix. Special enzymes and proteins induce torsional stresses into the molecule. The enzymes thereby alter the three-dimensional shape of DNA while maintaining its primary structure and the genetic information encoded in it. These enzymes are also essential for DNA replication and transcription. When a circular supercoiled DNA molecule is replicated, the two daughter molecules would become interlocked without a means of removing the supercoils and separation of the progeny would be impossible. There are, however, enzymes that remove the supercoils to allow the separation of daughter chromosomes, and subsequently catalyse supercoiling to facilitate efficient packaging of DNA. The so-called catenation of chromosomes can perhaps be more easily understood by performing the simple model experiment shown in Figure 4.12. Similarly when DNA is transcribed into RNA it is essential for the DNA to undergo localized swivelling movements otherwise

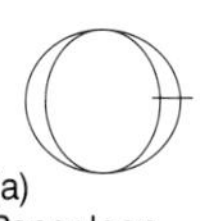

(a) Paper loop, introduce a nick by cutting

(b) Pass the strip through the nick

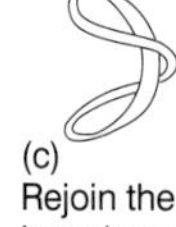

(c) Rejoin the loop (one supercoil)

(d) Replicate by cutting the loop lengthways

FIGURE 4.12 A schematic representation of the catenation of circular DNA molecules. The reader is encouraged to try this experiment. Take a strip of paper (a), introduce one complete twist into it (one supercoil) by passing the strip through a break (b) and rejoining (c). Cut down the length of the strip as if in semi-conservative replication. The result is two 'daughter molecules' which are interlinked (d). If two supercoils are introduced then a more complex linkage can be seen.

the RNA polymerase and the growing RNA chain would have to revolve around the DNA template.

The enzymes that facilitate these topological changes are known as topoisomerases, four types of which have been identified in *Escherichia coli*. The type II isomerase, or DNA gyrase, is the only one of the four that can introduce negative supercoils. It consists of two 97 kDa A chains and two 90 kDa B chains and is the target for several classes of antibacterial drugs.

The supercoiling reaction begins with a segment of DNA, approximately 120 bp in length, wrapping itself around the tetrameric complex of the gyrase. The B subunit allows passage of the DNA segment into the interior of the enzyme where both strands of DNA are cleaved by the A subunit, using enzyme-bound ATP as an energy source. The 5′-phosphate terminus of each cleaved strand is covalently bonded to the hydroxyl group of tyrosine-122 in each of the A subunits. This linkage is essential to prevent the free rotation of the cut DNA strands. The enzyme next permits a segment of double-stranded DNA to pass through the gap of the broken sequence. The ends of the cleaved strands are then brought together and resealed by the ligase activity of the enzyme. Finally, the B subunit catalyses the hydrolysis of the bound ATP, permitting the release of the processed DNA segment.

Quinolones

These compounds comprise one of the most important groups of wholly synthetic antibacterial drugs in current medical use. The original compound in the series, nalidixic acid (Figure 4.13), is only useful against infections caused by Gram-negative bacteria. The introduction into the molecule of 6-fluoro and 5-piperidine groups greatly increased antibacterial potency, although the activity of compounds such as norfloxacin (Figure 4.13) against Gram-positives remains limited. Fortunately, several more recent fluoroquinolones currently under investigation show greater promise against Gram-positive pathogens.

The antibacterial activity of the quinolones is primarily due to inhibition of DNA gyrase. When the isolated enzyme is incubated with DNA and a quinolone the supercoiling reaction is arrested at the point when the cut ends of the DNA strands are covalently linked to the hydroxyl groups of the tyrosine-122 residues of the A subunits. The re-ligation of the broken strands is blocked and the supercoiling reaction can be said to have been 'frozen' midway. This results in the accumulation of double-stranded nicks in the bacterial genome and may also prevent the essential movements of DNA and RNA polymerases along the DNA template. The bactericidal action of the quinolones probably arises from a combination of these effects.

Studies with DNA gyrase from quinolone-resistant bacteria first demonstrated that the A subunits are the target for quinolone action. The molecular details of the interaction between quinolones and the enzyme remain unclear. However, certain

Nalidixic acid

Norfloxacin

FIGURE 4.13 Two examples of quinolone antibacterial drugs that inhibit the A subunit of DNA gyrase.

amino acid residues lying between alanine-67 and glutamine-106 appear to be important in the binding. Serine-83 is especially significant because replacement of this amino acid by tryptophan markedly reduces quinolone binding. One current proposal is that the drug enters a binding pocket in the A subunit created by partial unwinding of DNA. The pocket is thought to contain four drug molecules held together by 'stacking' interactions between their planar ring systems. In one model of the drug–enzyme interaction, hydrogen bonding between the 3-carboxyl and 4-oxo groups of the quinolone molecule to adjacent bases of DNA contributes to the stability of the complex. However, another model suggests that binding between the drug and enzyme is more likely to involve the 3-carboxyl and 4-oxo groups with hydogen bond donors of key amino acids in the A subunit, such as serine-83 and glutamine-106. Clearly, more research, particularly X-ray crystallographic data, is needed to clarify the nature of the interaction between the quinolones and DNA gyrase.

Although the type II topoisomerase is believed to be the major target for quinolones, recent evidence indicates that the type IV enzyme is also inhibited by some fluoroquinolones, although to a lesser extent than the type II enzyme. Unlike DNA gyrase, type IV topoisomerase does not induce negative supercoiling of DNA but plays an important role in the partitioning of DNA during cell division. Its principal catalytic function is to ensure the decatenation of DNA that is essential for the correct separation of DNA into the daughter cells. Despite the distinct functions of the two enzymes, there is extensive amino acid sequence homology between them, which may underlie their susceptibility to inhibition by quinolones. Evidence that inhibition of topoisomerase IV may contribute to the antibacterial activity of fluoroquinolones came from studies with bacterial mutants resistant to high concentrations of these drugs. Further mutations in the genes mediating resistance to both gyrase and topoisomerase IV are necessary for high-level drug resistance.

Inhibitors of the B subunits of DNA gyrase: coumarins and cyclothialidines

The coumarins, exemplified by novobiocin (Figure 4.14) and the cyclothialidines (e.g. GR122222X; Figure 4.14), are structurally distinct families of naturally occurring inhibitors of DNA negative supercoiling; their target is the ATP hydrolysing activity of the B subunit.

Studies on the kinetics of inhibition of DNA gyrase by both types of compound indicate potent competition with ATP, with inhibitor constant (K_i) values in the range 10^{-7}–10^{-9} M. However, examination of the ATPase activity of isolated B subunits indicated that it did not follow Michaelis–Menten kinetics, thus raising doubts about the concept of competitive inhibition. These doubts were supported by the absence of any significant structural similarity between the inhibitors and ATP. Furthermore, point mutations in the B subunit that cause resistance to the coumarins were found to lie at the periphery of the ATP binding site. Fortunately, the uncertainty was eventually

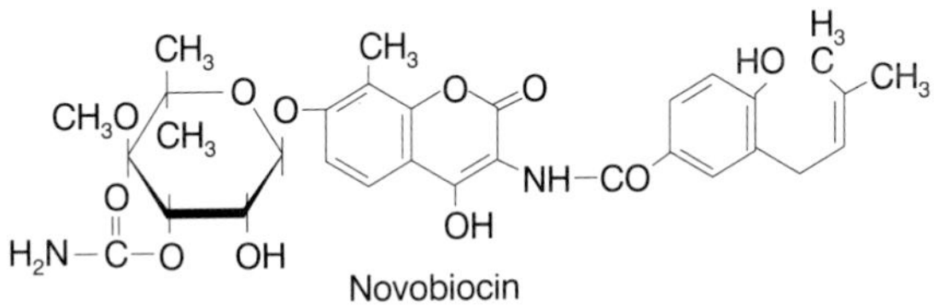

Novobiocin

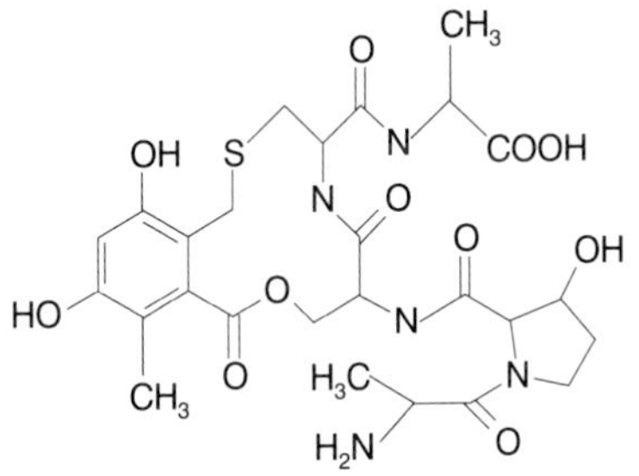

Cyclothialidine GR122222X

FIGURE 4.14 Naturally occurring (novobiocin) and synthetic (cyclothialidine) inhibitors of the function of the B subunit of bacterial DNA gyrase.

resolved by X-ray crystallographic data on the complexes of the B subunit with novobiocin, GR122222X and adenylyl-β-γ-imidodiphosphate (an analogue of ATP). Although the three ligands are structurally distinct and bind to the enzyme in very different ways, the X-ray analysis shows that there is some overlap of their binding sites. Taken with the convincing enzyme kinetics data there is therefore strong evidence for the competitive inhibition of ATP binding to the B subunit by both the coumarins and cyclothialidines.

As would be expected, the interactions of novobiocin and GR122222X with the enzyme are complex and the reader is referred to the research paper listed under 'Further reading' for detailed information. To summarize, in the case of novobiocin the novobiose sugar unit contributes to extensive hydrogen bonding with key amino acids of the enzyme and also with water molecules associated with the protein. Hydrophobic interactions involving both the sugar and coumarin elements add to the stability of the complex. The binding contribution of the 3′-iso-pentenyl-4′-hydroxybenzoate group is apparently the least significant because a chemically modified derivative of novobiocin lacking this group still inhibits both DNA supercoiling and the ATPase activity. The antibacterial activity of this compound is, however, much reduced and it is possible that the 3′-iso-pentenyl-4′-hydroxybenzoate group facilitates the entry of novobiocin into bacterial cells.

The binding of GR122222X to the B subunit is stabilized largely by the formation of water-mediated hydrogen bonds between the hydroxyl groups of the substituted resorcinol group of the inhibitor and key amino acids that line the binding site.

The clinical history of novobiocin and other coumarins has been disappointing, due to several factors. The coumarins are poorly absorbed after oral dosing and their antibacterial activity is limited mainly to Gram-positive organisms, probably because of limited penetration into Gram-negative cells. Drug-resistant mutants emerge readily from initially drug-sensitive populations of Gram-positive bacteria. Finally, although there is no functional equivalent of DNA gyrase in mammalian cells, mammalian topoisomerase II, which shares some sequence homology with gyrase, is susceptible to inhibition by novobiocin. In contrast, GR122222X is much more specific for the bacterial enzyme and the cyclothialidines may offer better prospects for clinical success than the coumarins.

Further reading

Gootz, T. D. and Brighty, K. E. (1996). Fluoroquinolone antibacterials: SAR, mechanism of action, resistance and clinical aspects. *Medicin. Res. Rev.* **16**, 433.

Hitchings, G. H. (1983). Inhibition of folate metabolism in chemotherapy. *Handb. Exp. Pharmacol.* **64**, 11.

Katz, R. A. and Skalka , A. M. (1994). The retroviral enzymes. *Ann. Rev. Biochem.* **63**, 133.

Kumar, K. P. Reddy, P. S. and Chatterji, D. (1992). Proximity relationship between the active site of *Escherichia coli* RNA polymerase and rifampin binding domain: a resonance energy-transfer study. *Biochemistry* **31**, 7519.

Lewis, R. J. *et al.* (1996). The nature of inhibition of DNA gyrase by the coumarins and the cyclothialidines revealed by X-ray crystallography. *EMBO J.* **15**, 1412.

Prescott, L. M., Harley, J. P. and Klein, D. A. (1996) *Microbiology*, 3rd edn, William C. Brown.

Richman, D. D. (1994). Drug resistance in viruses. *Trends Microbiol.* **2**, 401.

Wilson, W. D. and Jones, R. L. (1981). Intercalating drugs: DNA binding and molecular pharmacology. *Adv. Pharmacol. Chemother.* **18**, 177.

Chapter five

Inhibitors of protein biosynthesis

The process of protein synthesis, in which the information encoded by the four-letter alphabet of nucleic acid bases is translated into defined sequences of amino acids linked by peptide bonds, is an exquisitely complex process involving more than 100 macromolecules. Amino-acid-specific transfer RNA (tRNA) molecules, messenger RNAs (mRNAs) and many soluble proteins are required, in addition to the numerous proteins and three types of RNA that comprise the ribosomes. Although many general features of the protein synthetic machinery are similar in prokaryotic and eukaryotic organisms, a number of naturally occurring compounds specifically inhibit bacterial protein synthesis and thereby provide us with therapeutic agents of considerable value. Intriguingly, 'nature' has been much more successful in producing compounds that discriminate between bacterial and mammalian protein synthesis than synthetic organic chemistry. No wholly synthetic compound capable of specific inhibition of bacterial protein synthesis has yet emerged from organic chemistry laboratories. Inhibitors specific for protein synthesis in fungi are as yet unknown, probably because the similarities between the mechanisms in fungal and mammalian cells are too close to permit this degree of discrimination.

We provide a brief outline of the process of protein synthesis and the main points of difference between that in bacterial and mammalian cells, and then go on to consider the actions of several inhibitors of bacterial protein synthesis.

5.1 Ribosomes

These remarkable organelles are the machines upon which polypeptides are elaborated. There are three main classes of ribosomes, identified by their sedimentation coefficients in the ultracentrifuge. The 80S ribosomes are apparently confined to eukaryotic cells, while 70S ribosomes are characteristic of prokaryotic cells. A species of 50–55S ribosome found in mammalian mitochondria resembles bacterial ribosomes in functional organization and antibiotic sensitivity; analogous small ribosomes also occur in the chloroplasts of green plants. The 80S particle dissociates reversibly into 60S and 40S subunits, and the 70S ribosome into 50S and 30S subunits, when the Mg^{2+} concentration of a suspending solution is reduced. Both 80S and 70S ribosomes are composed exclusively of protein and RNA in mass ratios of approximately 50:50 and 35:65, respectively. There are three distinct species of RNA in most ribosomes, with sedimentation coefficients of 29S, 18S and 5S in 80S particles from animal cells; 25S, 18S and 5S in 80S particles from plant cells; and 23S, 16S and 5S in 70S particles. The 55S ribosomes contain two RNA species that sediment at about 16S and 12S, but probably not 5S RNA. The protein composition of ribosomes is impressively complex. The 30S subunit of *Escherichia coli* ribosomes contains 21 proteins ('S' proteins), and the 50S subunit 34 proteins ('L' proteins). The amino acid sequences of all of these proteins are now known.

Great progress has been made in understanding how the ribosome is constructed and the leading groups in the field have agreed on a consensus model of the structure of 70S ribosomes, a simplified representation of which is shown in Figure 5.1. Two main functional regions have been defined, known as the translational and exit domains. Both ribosomal subunits contribute to the translational domain; mRNA binding and its interaction with aminoacyl-tRNAs, i.e. decoding, occur on the 30S subunit in the region called the platform; while peptide bond formation is catalysed by the peptidyl transferase activity located on the central protuberance of the 50S subunit. The lengthening peptide chain leaves the 50S subunit via the exit domain which is found on the side of the subunit opposite to the peptidyl transferase site. The topographies of the various elongation and initiation factor binding sites are also well understood and represent the culmination of many years of effort by a number of groups using highly sophisticated techniques. Similar features have been recognized in eukaryotic ribosomes, which bind to the rough endoplasmic reticulum at the exit site.

Recently there has been something of a revolution in the assessment of the relative contributions of ribosomal proteins and ribosomal RNA to protein synthesis. For many years it was believed that ribosomal proteins played the leading role in decoding mRNA and in peptide bond synthesis. However, the weight of evidence now points to ribosomal RNA being critically involved both in decoding (16S rRNA) and peptide bond synthesis (23S rRNA) with certain ribosomal proteins being required to maintain the functional three-dimensional structures of the RNA molecules. As we shall see, these discoveries have major implications for the sites of action of several antibiotics that inhibit protein biosynthesis.

5.2 Stages in protein biosynthesis

5.2.1 Formation of aminoacyl-transfer RNA

Each amino acid is converted by a specific aminoacyl-tRNA synthetase to an aminoacyladenylate which is stabilized by association with the enzyme:

$$\text{ATP} + \text{amino acid (aa)} \xrightleftharpoons{\textit{Aminoacyl-tRNA synthetase}} \text{aa-AMP-Enz} + \text{PP}_i.$$

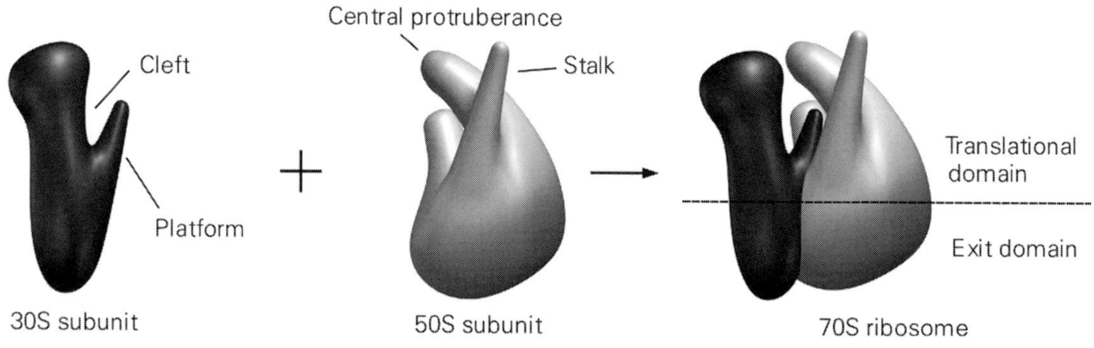

FIGURE 5.1 Simplified representation of the subunits of a prokaryotic ribosome and their co-operative interaction to form the functional 70S particle. This diagram was kindly provided by Paul J. Franklin.

Each amino acid–adenylate–enzyme complex then interacts with an amino-acid-specific tRNA to form an aminoacyl-tRNA in which the aminoacyl group is linked to the 3'-OH ribosyl moiety of the 3' terminal adenosyl group of the tRNA by a highly reactive ester bond.

$$\text{aa-AMP-Enz} + \text{tRNA} \leftrightarrow \text{aminoacyl-tRNA} + \text{AMP} + \text{Enz}$$

The subsequent stages in prokaryotic protein biosynthesis are outlined in Figure 5.2

5.2.2 Initiation

The mechanism of initiation has been analysed in detail. Three protein factors, IF1, IF2 and IF3, loosely associated with 70S ribosomes are concerned with initiation. IF1 enhances the rate of ternary complex formation between mRNA, initiator tRNA and 30S ribosomal subunits. IF1 also has a role in promoting the dissociation of 70S ribosomes released from previous rounds of polypeptide synthesis into 30S and 50S subunits. Factor IF3, which then binds to the 30S subunit, is also needed for the binding of mRNA. The complex containing the 30S subunit, IF3 and mRNA is joined by IF2, GTP and the specific initiator tRNA, *N*-formylmethionyl-tRNA$_F$ (fMet-tRNA$_F$), the role of IF2 being to direct the binding of fMet-tRNA$_F$ to a specific initiator codon, usually AUG but occasionally GUG. IF1 and IF2 are now ejected from the complex, a process dependent on the hydrolysis of one molecule of GTP to GDP and inorganic phosphate. The next stage involves the detachment of IF3 in the presence of a 50S subunit to permit the formation of the 70S ribosome. The association of the 50S and 30S subunits is believed to involve interactions between both their protein and RNA chains.

Initiation on 80S ribosomes is thought to resemble that on 70S ribosomes except that eukaryotic initiation *in vivo* uses unformylated Met-tRNA^{Met}. In addition, there are at least nine eukaryotic initiation factors whose interplay is much more complex than that in prokaryotic organisms.

5.2.3 Peptide bond synthesis and chain elongation

The consensus view of synthetic sequence rests largely on the concept of two distinct sites on the ribosome, called the acceptor (A) site and the donor or peptidyl (P) site. The A site is the primary decoding site where the codon of the mRNA first interacts with the anticodon region of the specific aminoacyl-tRNA. The fMet-tRNA$_F$, however, binds directly to the P site. The binding of the next aminoacyl-tRNA to the A site requires protein factors EF-T$_s$ and EF-T$_u$. EF-T$_u$ binds GTP and then forms a ternary complex with aminoacyl-tRNA. This complex binds to the acceptor site, with accompanying hydrolysis of one molecule of GTP. GTP hydrolysis is not essential for the binding of aminoacyl-tRNA, but in its absence the bound aminoacyl-tRNA is not available for peptide bond formation. The role of the stable factor, EF-T$_s$, is to regenerate EF-T$_u$-GTP from EF-T$_u$-GDP by stimulating the exchange of bound GDP for a molecule of free GTP. Apparently EF-T$_s$ forms a high-affinity intermediate complex with EF-T$_u$ and GDP is lost from this intermediate.

The scene is now set for the formation of the first peptide bond. The carboxyl group of the *N*-formylmethionine attached to the P site through its tRNA is 'donated' to the amino group of the adjacent amino acid at the A site, to form a peptide bond. The formation of the peptide bond is catalysed by peptidyl transferase, which is a complex component of the 50S subunit located at the central protuberance of the 50S subunit. Although several ribosomal proteins, including L2, L15, L16 and L27, are located in the vicinity of the peptidyl transferase site, the current view is that 23S rRNA is responsible for catalysing peptide bond formation. The reader is referred to the relevant papers listed under 'Further reading' for further details of

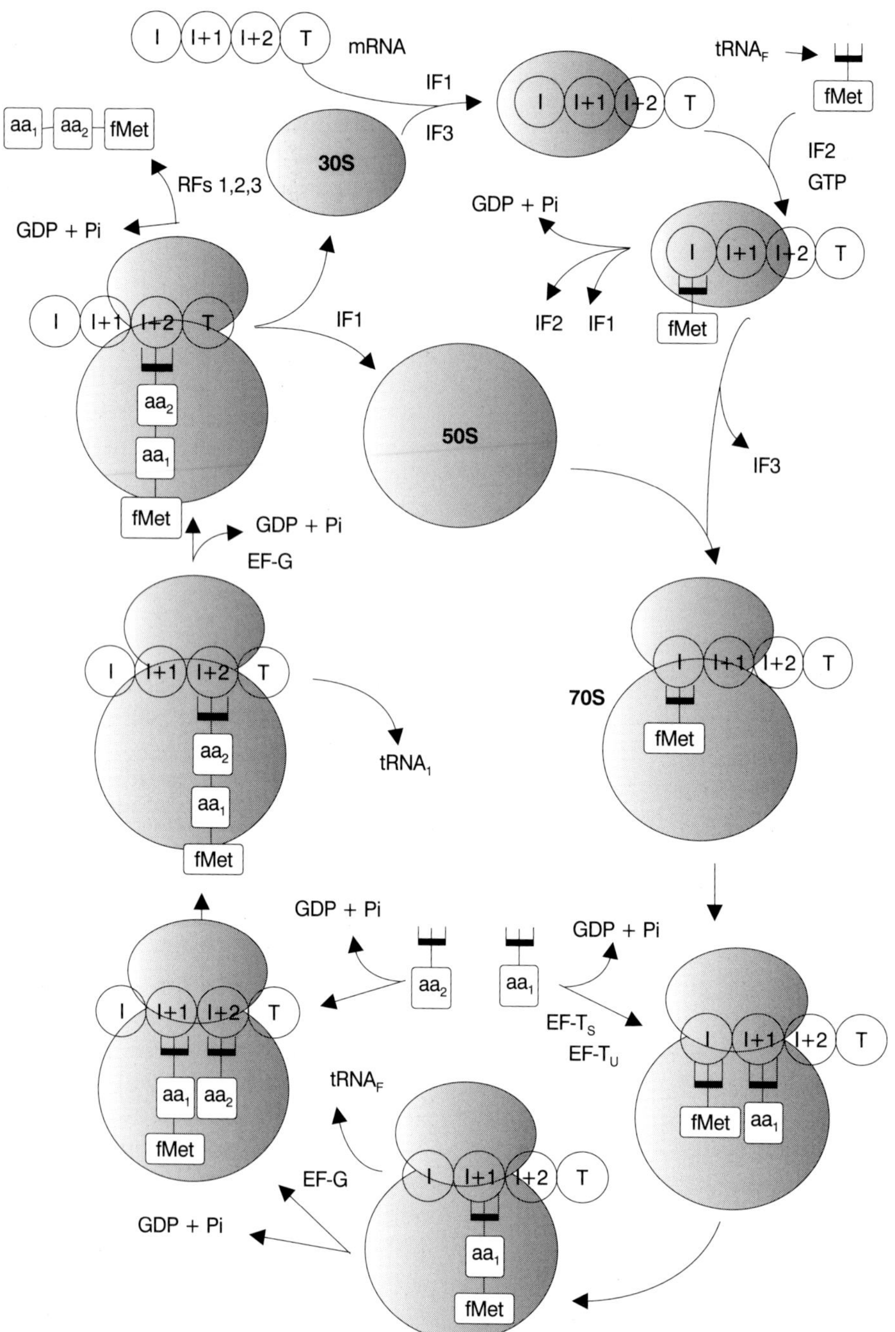

FIGURE 5.2 Diagrammatic summary of the main stages in the biosynthesis of proteins on 70S ribosomes. The scheme should be read clockwise starting at the top. I, I+1, I+2 represent the initiator and successive codons; T, the terminator codon on mRNA; fMet, aa_1, aa_2, *N*-formylmethione and two other amino acids; $tRNA_F$, $tRNA_1$, $tRNA_2$, specific transfer RNAs. The involvement of the various protein cofactors referred to in the text is also indicated.

the complex structures of both 16S and 23S rRNA. Suffice it to say that the peptidyl transferase function is probably located in the region of the 23S rRNA known as domain V, which also harbours the binding site for protein L27. The dipeptide which is formed at domain V remains attached through its C-terminus to the second tRNA at the A site. The dipeptidyl tRNA is then translocated from the A to the P site, still linked to the mRNA (through the codon–anticodon interaction). The third consecutive codon of the mRNA is now exposed at the A site by the relative movement of the ribosome towards the 3′ end of the mRNA. The translocation step requires factor EF-G and the hydrolysis of another molecule of GTP. EF-G binds to the L7/L12 region of the large subunit. This area is implicated in GTP hydrolysis mediated by EF-T_u, EF-G and by IF2.

5.2.4 Chain termination and release

The signal for termination of the polypeptide chain is given by the appearance of one of three terminator codons, UAA, UAG or UGA, at the A site. The complete polypeptide is detached from the tRNA at the C-terminal amino acid, a step that requires peptidyl transferase activity and the release factors RFl, RF2 and RF3. Factors RFl and RF2 are concerned with the recognition of specific terminator codons, RFl recognizing UAA and UAG and RF2, UAA and UGA. Both the binding of RFl and RF2 to the ribosomes and their release require RF3. Cleavage of GTP is also involved in the release reaction. Release from eukaryotic ribosomes involves only one codon-recognizing release factor and also requires the hydrolysis of GTP. The formyl groups of the fMet ends of prokaryotic polypeptides are removed by a specific enzyme and in many proteins the methionine residue is also removed. After release of the completed polypeptide, the ribosome is liberated from the mRNA and deacylated tRNA by the action of GTP, EF-G and ribosome release factor (RRF) which permits dissociation into 30S and 50S subunits. IF3 then binds to the 30S subunit. This prevents reassociation until a full initiation complex has once more been completed.

5.3 Puromycin

The antibiotic puromycin is a unique inhibitor of protein biosynthesis, since the drug itself reacts to form a peptide with the C-terminus of the growing peptide chain on the ribosome, thus prematurely terminating the chain. This remarkable property of puromycin gave the antibiotic an important role in the elucidation of the mechanism of peptide bond formation and, as we shall see, in defining the point of action of several other inhibitors of protein biosynthesis.

The structural similarity of puromycin to the terminal aminoacyladenosine moiety of tRNA was noted many years ago (Figure 5.3) and this proved to be the key to understanding its actions. Since aminoacyladenosine is the terminal residue of tRNA in both prokaryotic and eukaryotic organisms, puromycin terminates protein synthesis equally effectively on 70S and 80S ribosomes and therefore has no therapeutic value. The structural analogy of puromycin with aminoacyladenosine led to the demonstration that the amino group of the antibiotic forms a peptide bond with the acyl group of the terminal aminoacyladenosine moiety of peptidyl-tRNA attached to the ribosome. No further peptide bond formation is possible because of the chemical stability of the C–N bond which links the *p*-methoxyphenylalanine moiety of puromycin to the nucleoside residue. Peptidyl-puromycin then dissociates from the ribosome.

Provided that the peptidyl-tRNA is in the P site on the ribosome, its reaction with puromycin has no other requirement than a normally functioning peptidyl transferase activity. Puromycin does not, however, react with peptidyl-tRNA in the A site; in this situation factor EF-G and GTP must be added in order to effect translocation of the peptidyl-tRNA to the P site. Only then is peptidyl-puromycin formed and released from the ribosome. The puromycin reaction occurs fairly readily at 0 °C, whereas normal chain elongation is negligible at this

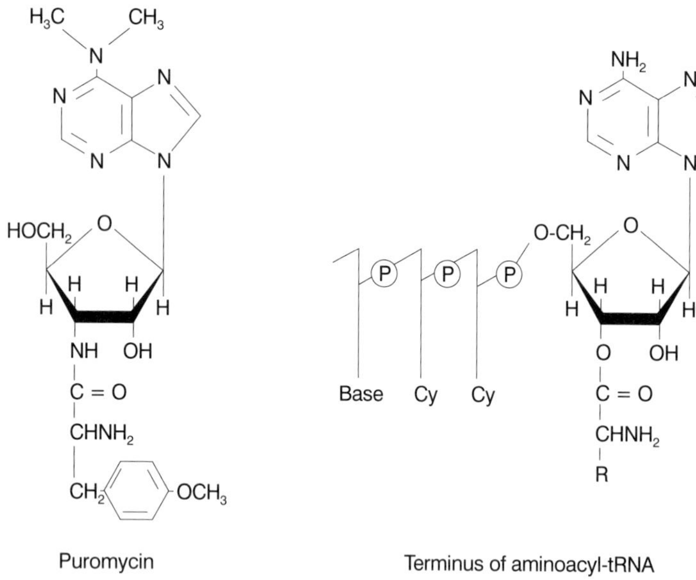

Puromycin

Terminus of aminoacyl-tRNA

FIGURE 5.3 Structural analogy between puromycin and the aminoacyl terminus of transfer RNA. Cy represents cytosine, and R the rest of the amino acid molecule.

temperature, suggesting that puromycin has a considerable competitive advantage over aminoacyl-tRNA in reacting with the peptidyl-tRNA. The puromycin reaction will, in fact, proceed under greatly simplified conditions requiring only 50S ribosomal subunits, the oligonucleotide CAACCA-(fMet) to replace the peptidyl tRNA normally found at the P site, Mg^{2+} and K^{+} ions. This simple system, known as the fragment reaction, allows the separation of peptide bond formation from the much more complex process of translation. It has been extremely useful in the investigation of those antibiotics suspected of inhibiting peptide bond synthesis.

Derivatives of puromycin indicate that a single benzene ring in the side chain is necessary for activity; replacement of the *p*-methoxyphenylalanine with proline, tryptophan, benzylhistidine or any aliphatic amino acid results in a marked loss of activity. The L-phenylalanine analogue is about half as active as puromycin, while the D-phenylalanine analogue is completely inactive. Replacing the *p*-methoxyphenylalanine residue with the *S*-benzyl-L-cysteine analogue results in only a minor loss of activity, which may be due to the increased distance between the benzene ring and the free NH_2 group caused by the additional S and C atoms. Since puromycin substitutes for all aminoacyl-tRNAs equally well, the sufficiency of a single benzene ring in the amino acid moiety of puromycin and its analogues is puzzling.

The aromatic ring may be involved in a hydrophobic interaction with the terminal adenosine of peptidyl-tRNA at the donor site, thus contributing to the formation of an intimate complex between puromycin and peptidyl-tRNA prior to the formation of a peptide bond. In view of the structure of the aminoacyladenosine of the tRNA terminus, the requirement for linkage of the amino acid moiety to the ribose 3′ position of puromycin is, however, not unexpected. Puromycin substituted in the 5′ position of the ribose with cytidylic acid is an effective peptide chain terminator, and there is an absolute requirement for cytidine in this derivative. Presumably this substitution extends the structural analogy with tRNA.

5.4 Inhibitors of aminoacyl-tRNA formation

A number of naturally occurring and synthetic analogues of amino acids inhibit the formation of the aminoacyl-tRNA complex. Close analogues may become attached to the appropriate tRNA and subsequently become incorporated into abnormal proteins. Among these are ethionine, norleucine, *N*-ethylglycine and 3,4-dehydroproline. Several naturally occurring antibiotics, such as borrelidin, furanomycin and indolmycin, competitively antagonize the incorporation of the corresponding amino acids, i.e. threonine, isoleucine and tryptophan respectively, into aminoacyl-tRNA. Most of these inhibitors of aminoacyl-tRNA formation lack species specificity and hence have no useful medical application, although indolmycin is said to be specific for prokaryotic tryptophanyl-tRNA synthetase. However, by far the most important inhibitor of aminoacyl-tRNA synthesis is the antibiotic mupirocin or pseudomonic acid A (Figure 5.4), which is produced by *Pseudomonas fluorescens*. Mupirocin has excellent activity against several species of *Staphylococcus*, and is especially useful against the dangerous multiply drug-resistant *Staphylococcus aureus* (MRSA). The antibiotic has limited activity against Gram-negative bacteria but includes in its spectrum *Haemophilus influenzae, Neisseria gonorrhoeae* and *Neisseria meningitidis.* Unfortunately, although mupirocin is well absorbed after oral dosing it is rapidly metabolized in the body to the inactive monic acid. Its clinical use is therefore confined to topical application, for example to eliminate MRSA from the nasal passages of hospital staff and vulnerable patients.

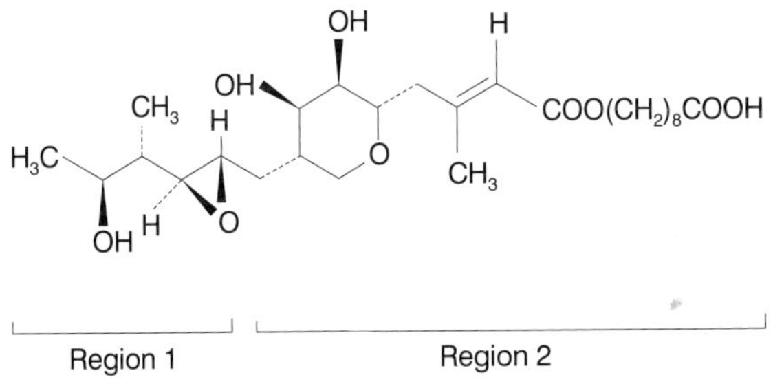

FIGURE 5.4 Mupirocin (pseudomonic acid A).This antibiotic, which has useful topical activity against multiply drug-resistant *Staphylococcus aureus,* is a specific inhibitor of bacterial isoleucyl-tRNA synthetase.

The antibacterial activity of mupirocin depends upon its specific inhibition of isoleucyl-tRNA synthetase. The compound is a competitive inhibitor of the isoleucyl substrate, probably because of a resemblance between the carbon skeleton of the terminus of the antibiotic which contains the epoxide residue (region 1, Figure 5.4) and that of isoleucine. Recent work suggests that the remainder of the mupirocin molecule (region 2) may bind to the ATP-binding subsite in the enzyme. The antibiotic can therefore be regarded as a bifunctional inhibitor of isoleucyl-tRNA synthetase, having structural features resembling both isoleucine and ATP. Region 1 of mupirocin is believed to bind to the enzyme in the vicinity of the amino acid sequence PYVPGWCHGL, and region 2 close to the KMSKS sequence. The antibacterial specificity of mupirocin rests on the fact that the compound has little or no inhibitory activity against eukaryotic isoleucyl-tRNA synthetase.

5.5 Inhibitors of initiation and translation

5.5.1 Streptomycin

This naturally occurring antibiotic is a member of the aminoglycoside group and has the complex chemical structure illustrated in Figure 5.5. While there is considerable variation in the chemical structures of different aminoglycosides (Figure 5.6), they all possess a cyclohexane ring bearing basic groups in the 1 and 3 positions with oxygen substituents at 4, 5 and 6 and sometimes at 2. All these groups are in equatorial positions.

Streptomycin was discovered by Waksman in the early 1940s and it was the first really effective drug against tuberculosis, although it is little used in the treatment of this disease today. It is a broad-spectrum antibiotic, active against a range of Gram-positive and Gram-negative bacteria, but its use is limited by several problems. First, the drug is only effective when given by injection because its absorption from the gastrointestinal tract is very poor. Secondly, along with other aminoglycosides, streptomycin may cause permanent deafness

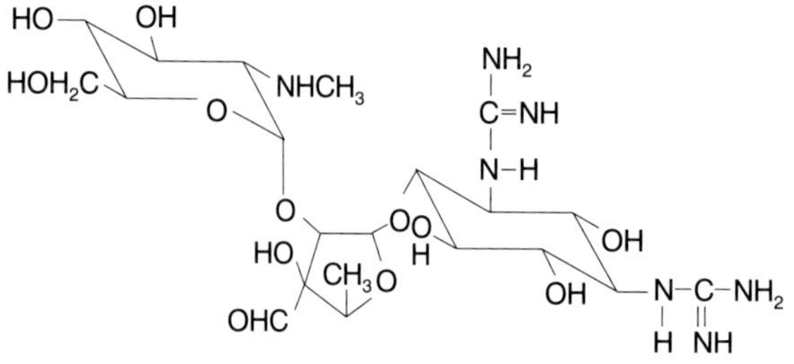

FIGURE 5.5 The aminoglycoside streptomycin, the first effective antitubercular drug.

due to irreversible injury to the eighth cranial nerve and also kidney damage, although fortunately the latter is usually reversible. Thirdly, bacterial resistance develops readily against this antibiotic.

Streptomycin is bactericidal, but cell death is preceded by marked effects on protein biosynthesis which are specific for the 70S ribosomes of bacteria:

(1) Streptomycin strongly inhibits the initiation of peptide chains. The drug also slows the elongation of partly completed chains, although even at high concentrations of streptomycin chain elongation is not completely suppressed. Peptidyl transferase activity is unaffected. These effects on initiation and elongation are attributed to a disturbance of the functions of both A and P sites by streptomycin.

(2) Many studies have been carried out on the effects of streptomycin in cell-free systems using synthetic polynucleotides as messengers, and some strikingly varied effects are observed depending on the mRNA used. Thus strepto-

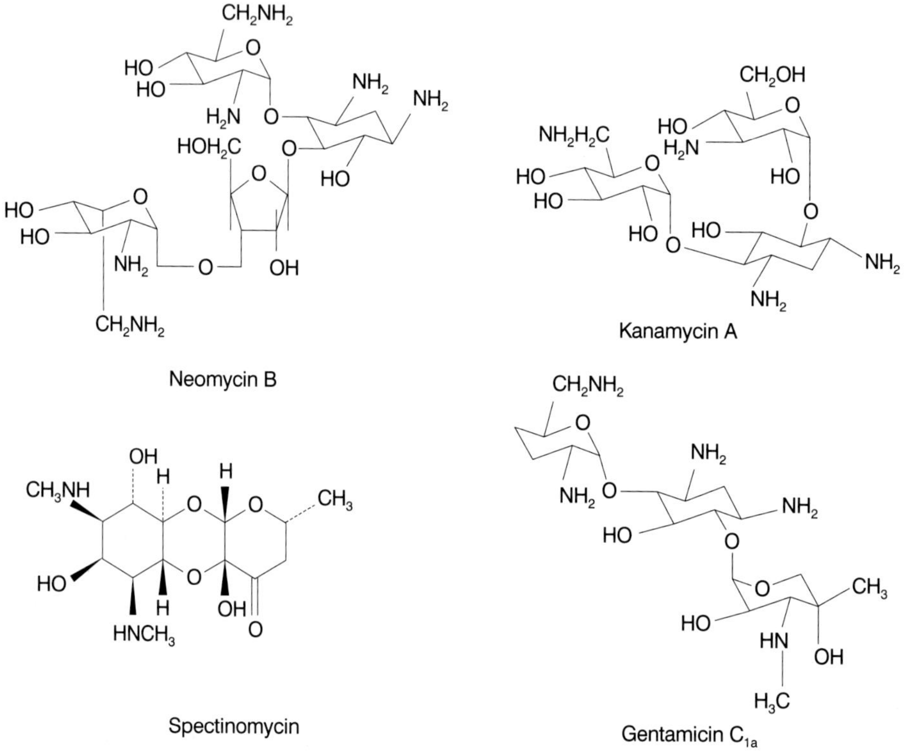

Neomycin B

Kanamycin A

Spectinomycin

Gentamicin C_{1a}

FIGURE 5.6 Examples of other aminoglycoside antibiotics. Spectinomycin is more appropriately described as an aminocyclitol antibiotic as it contains an inositol ring with two of its OH groups substituted by methylamino groups.

mycin inhibits the incorporation into peptide linkages of:

(a) phenylalanine directed by poly(U);
(b) histidine and threonine directed by poly(AC); and
(c) arginine and glutamic acid directed by poly(AG).

In contrast, streptomycin may, under some conditions, stimulate the incorporation of amino acids in the presence of synthetic messengers which do not normally code for these amino acids. For example, while streptomycin inhibits the incorporation of phenylalanine in the presence of poly(U), the incorporation of isoleucine and serine is stimulated. Streptomycin also induces poly(C) to promote the incorporation of threonine and serine instead of proline. All these observations indicate that streptomycin distorts the proof-reading selection of the correct aminoacyl-tRNA by the ribosome. However, the misreading is not random and the following rules are more or less observed:

1. In any mRNA codon only one base is misread, usually a pyrimidine located at the 5′ end or middle position of the codon.
2. There is no misreading of the base at the 3′ end.
3. Misreading of purines is rare and the occurrence of these in a codon decreases the chance of misreading the codon.

The induction of misreading of the genetic message by streptomycin probably underlies the well-known ability of this antibiotic to suppress certain bacterial mutations.

Site of action of streptomycin

Streptomycin binds tightly but not irreversibly to 70S ribosomes, with a K_d of 10^{-7} M. There is also low-affinity binding ($K_d > 10^{-4}$ M), which is probably irrelevant to the action of streptomycin. For many years it was thought that a unique ribosomal binding site for streptomycin had been identified by experiments with bacterial mutants highly resistant to the antibiotic. Ribosomes prepared from streptomycin-sensitive and streptomycin-resistant *Escherichia coli* were dissociated into 30S and 50S subunits by lowering the Mg^{2+} concentration in the medium. A ribosomal subunit ‘cross-over’ experiment showed that reassociated 70S particles composed of 30S subunits from resistant cells and 50S subunits from sensitive cells were resistant to streptomycin. In the opposite cross, i.e. 30S subunits from sensitive cells and 50S subunits from resistant cells, the resulting 70S ribosomes were streptomycin sensitive. This indicated that the target site of streptomycin was on the 30S subunit, a view strengthened by the finding that radiolabelled streptomycin bound specifically to the 30S subunit but not to the 50S subunit of sensitive ribosomes. Streptomycin did not bind to the 30S subunit from resistant cells and did not induce misreadings of mRNA translated with resistant ribosomes.

It was then found that in resistant ribosomes protein S12 was mutated. Several resistant variants were isolated with various single amino acid replacements. Thus lysine-42 was replaced by asparagine, threonine or arginine, while in another mutant lysine-87 was replaced by arginine alone. However, studies with 30S subunits treated with protein extractants to remove the S12 protein showed that it is not essential for protein synthesis nor is it an absolute requirement for streptomycin binding.

As mentioned above, evidence has accumulated of a central role for rRNA in protein biosynthesis. This has been accompanied by an increasing awareness of the importance of the interactions between rRNA and several antibiotics and their ability to interfere with various stages in the biosynthetic process. The crucial technique used to reveal the significance of rRNA is that of ‘footprinting’. This detects the ability of antibiotics bound to ribosomes to protect the bases of specific nucleotides of 16S and 23S rRNA against chemical modification, usually alkylation, by reagents such as dimethyl sulphate. Such protection is considered to be evidence of specific interactions or binding between the antibiotics and functionally significant domains in rRNA which would otherwise by attacked by the chemically reactive reagent. When streptomycin binds to 70S ribosomes in the presence of dimethyl sulphate electrophoretic analysis of 16S rRNA following its hydrolysis to individual nucleotides shows that the antibiotic affords protection to the bases of nucleotides 911 to 915, although protection is incomplete even when the streptomycin–ribosome binding is fully saturated. This suggests that the protection afforded by streptomycin may be indirect rather than due to direct contact between the antibiotic and 16S rRNA. Possibly the drug binds elsewhere in the ribosome and in doing so induces a conformational change in 16S rRNA that hides the bases of nucleotides 911–915 from chemical attack. Interestingly, a single base change of cytosine to uracil at position 912 in the 16S rRNA of *Escherichia coli* yields streptomycin-resistant bacteria.

At present, the ribosomal binding site for streptomycin cannot be identified with certainty, although both the S12 protein and the 911–915 region of 16S rRNA must be involved in the mediation of the effects of streptomycin on mRNA decoding and on the initiation of peptide chains.

The bactericidal action of streptomycin

Streptomycin and structurally related aminoglycosides are unusual among inhibitors of protein biosynthesis in causing the death of bacteria. Most other inhibitors merely arrest bacterial growth, which resumes when the antibiotic is removed from the microbial environment. It seems likely that the ability of the aminoglycosides to induce ribosomal misreading of mRNA may be an important factor in their bactericidal action. Very probably the resulting aberrant proteins cause a variety of disordered activities within the cell, amongst the most important of which may be disruption of the normal functions of the cytoplasmic membrane and the outer membrane of Gram-negative bacteria, leading to irreversible changes in cellular metabolism. The tight binding of streptomycin to ribosomes may also contribute to the lethality of action, although the K_d of 10^{-7} M indicates that the binding is reversible. Conceivably, streptomycin induces a relatively stable conformational change in the 30S subunit which persists after the dissociation of the antibiotic. However, definitive evidence for the basis of streptomycin lethality is lacking.

5.5.2 Other aminoglycoside antibiotics

In addition to streptomycin there are several other aminoglycosides that are useful in antibacterial chemotherapy. These include neomycin, kanamycin, gentamicin (Figure 5.6), tobramycin, amikacin and netilmicin, the last two of which are semi-synthetic modifications of naturally occurring antibiotics designed to minimize enzymic inactivation by resistant bacteria (Chapter 9). Neomycin and kanamycin are usually restricted to topical use, while the remaining four compounds are given by injection for the treatment of serious Gram-negative infections.

Several of these antibiotics have effects on protein biosynthesis that are distinct from those of streptomycin. For example, gentamicin, kanamycin and neomycin exhibit three separate concentration-dependent effects on isolated ribosomes:

1. At concentrations below 2 $\mu g\ ml^{-1}$ there is strong inhibition of total protein synthesis associated with inhibition of the initiation step, but little induction of mRNA misreading.
2. Between 5 and 50 $\mu g\ ml^{-1}$ there is misreading, especially by reading through the termination signals. Protein synthesis may therefore actually increase through the accumulation of abnormally long polypeptides as the ribosomes continue past the end of one message and on to the next.
3. Higher antibiotic concentrations re-establish inhibition of protein synthesis.

Each ribosomal subunit has one strong binding site for kanamycin as well as a number of weak binding sites. Gentamicin and neomycin compete with kanamycin, suggesting similar binding domains which must be distinct from the streptomycin-binding region since there is no comparable inhibition with streptomycin. This conclusion is supported by footprinting studies with neomycin, kanamycin and gentamicin which indicate a different pattern of protection from that of streptomycin. Neomycin, kanamycin and gentamicin protect adenine (A)-1408, guanine (G)-1491 and G-1494 of 16S rRNA from chemical modification by dimethyl sulphate. These bases are located close to the decoding region of the 3′ end of 16S rRNA, and two of them (A-1408 and G-1494) are also protected from chemical probing by tRNA bound to the ribosomal A site. These data are consistent, therefore, with the miscoding effects of neomycin, kanamycin and gentamicin, although the detailed molecular mechanism remains obscure. The basis of the different concentration-dependent effects of the antibiotics is also uncertain but may be associated with progressive saturation of their ribosomal binding sites with increasing concentration.

Spectinomycin is usually included in the aminoglycoside group, even though it lacks an amino sugar residue (Figure 5.6). Unlike the previously mentioned aminoglycoside antibiotics, its action is bacteriostatic rather than bactericidal. The effects of spectinomycin on protein synthesis are also markedly different from those of the other aminoglycosides. While it inhibits protein synthesis in bacterial cells and in cell-free systems containing 70S ribosomes, spectinomycin does not induce misreading of mRNA. Spectinomycin may inhibit an initial translocation step as it has no effect on codon recognition, peptide bond formation, or chain termination and release. With the emergence of β-lactamase-producing *Neisseria gonorrhoeae,* spectinomycin is finding a useful clinical application in the treatment of gonococcal infections.

5.5.3 Tetracyclines

Four important members of this group are illustrated in Figure 5.7. The tetracyclines are broad-spectrum antibiotics which are also effective against rickettsial organisms, mycoplasmas and certain protozoa. The antibiotic (bacteriostatic) activity of the tetracyclines depends on direct inhibition of protein biosynthesis. Unlike most other therapeutically useful inhibitors of protein biosynthesis, the tetracyclines inhibit both 70S and 80S ribosomes, although 70S ribosomes are rather more sensitive. However, the tetracyclines are much more effective against protein synthesis in bacterial cells than against eukaryotic cells because of the ability of sensitive bacteria to concentrate tetracyclines within their cytoplasm (Chapter 7).

Studies of the effects of the tetracyclines on the tRNA–ribosome interaction show that they inhibit the binding of aminoacyl-tRNA to the A site on the ribosome but have little effect on binding to the P site except at high drug concentrations. The binding of fMet-$tRNA_F$ to the ribosome is about one-tenth as sensitive to tetracycline as the binding of other aminoacyl-tRNAs, since fMet-$tRNA_F$ binds to the P site rather than to the A site. The tetracyclines do not directly inhibit formation of the peptide bond or the translocation step except at high concentrations. They have no effect on the hydrolysis of GTP to GDP required for the functional binding of aminoacyl-tRNA to the A site. Possibly, the tetracyclines uncouple GTP hydrolysis from the binding reaction. Tetracyclines also inhibit peptide chain termination and release by blocking the binding of the release factors at stop codons in the A site. However, it is unlikely that the effects on termination and release contribute significantly to the antibacterial action of tetracyclines.

There is a single, moderately high-affinity binding site for tetracycline on the 30S subunit, with a K_d of approximately 1 μM. There are also many low-affinity sites which are not considered essential to the inhibitory action of the drug. Photoaffinity labelling studies with a photoreactive derivative of tetracycline revealed extensive labelling of protein S7, which is located near the region of contact between the two ribosomal subunits. Footprinting studies failed to implicate 16S rRNA as an essential component of the binding site for tetracyclines. While tetracycline protected A-892 from alkylation by dimethyl sulphate, minocycline and doxycyline, which also inhibit ribosomal function in the same way as tetracycline,

Tetracycline

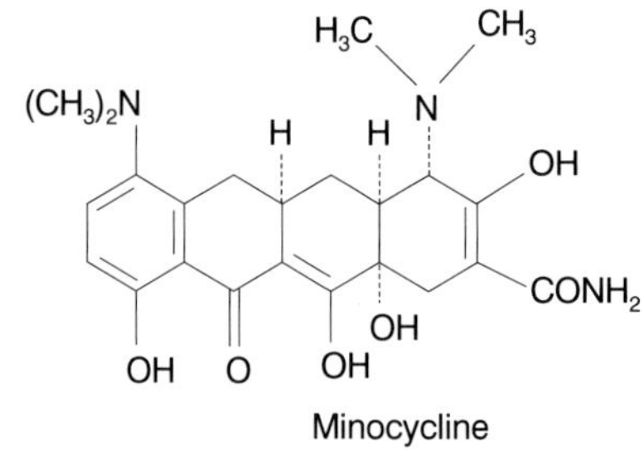

Oxytetracycline

Doxycycline

FIGURE 5.7 Four major tetracycline broad-spectrum antibiotics. Minocycline is effective against some bacteria that are resistant to the other drugs.

failed to do so. Thus it cannot be concluded that A-892 is central to the inhibition of ribosome function by the tetracycline group as a whole, although it may contribute to the binding of tetracycline itself.

The features of the tetracycline molecule for antibacterial activity have been worked out in some detail. If there is a specific permeation mechanism for the entry of tetracyclines into bacterial cells (Chapter 7), it may have its own structural requirements. It should not be assumed therefore that the structural features of the tetracycline molecule required for antibacterial activity are the same as those for the inhibition of ribosomal function. It is possible, therefore, that some tetracycline derivatives which inhibit ribosomes may lack antibacterial activity because of a failure to achieve an inhibitory concentration within the bacterial cell.

The more limited investigations on the structural requirements for the inhibition of protein synthesis on isolated ribosomes reveal several modifications in structure (Figure 5.8) that significantly affect inhibitory activity:

1. Chlorination of the 7 position significantly increases inhibitory activity.
2. Epimerization of the 4-dimethylamino group significantly decreases activity.
3. Both 4α,12α-anhydro- and 5α,6-anhydrotetracyclines are much less active than tetracycline.
4. Ring opening of the tetracycline nucleus to give the iso derivatives and the α and β isomers of apo-oxytetracycline destroys activity.
5. Replacement of the amidic function at C_2 with a nitrile group results in a marked loss of potency.

The ability of the tetracyclines to chelate polyvalent cations may have a bearing on their inhibition of protein biosynthesis. It is possible that Mg^{2+} ions attached to the phosphate groups of the ribosomal RNA may facilitate an initial interaction between the ribosome and tetracycline molecules. Free Mg^{2+} ions in the bacterial cytoplasm may also complex with tetracycline, thus limiting its ability to interact with ribosomal Mg^{2+}. The 11,12,β-diketone system, the 12α- and 3-hydroxyl groups have

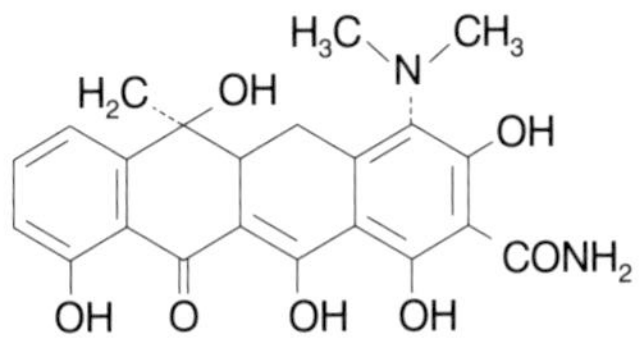

4α, 12α-Anhydrotetracycline

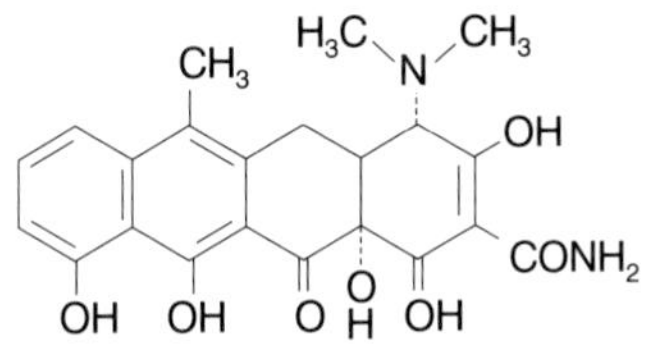

5α, 6-Anhydrotetracycline

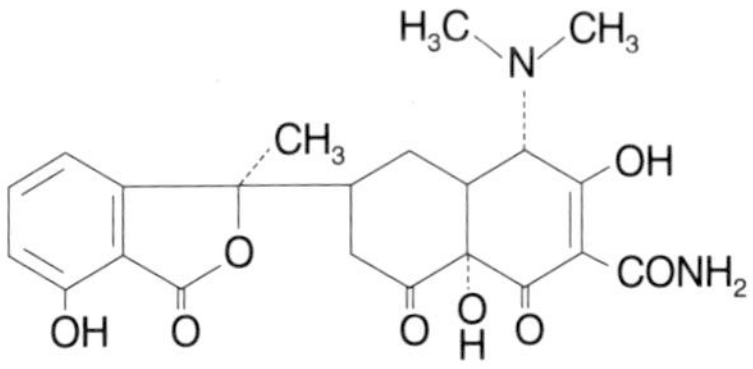

Isotetracycline

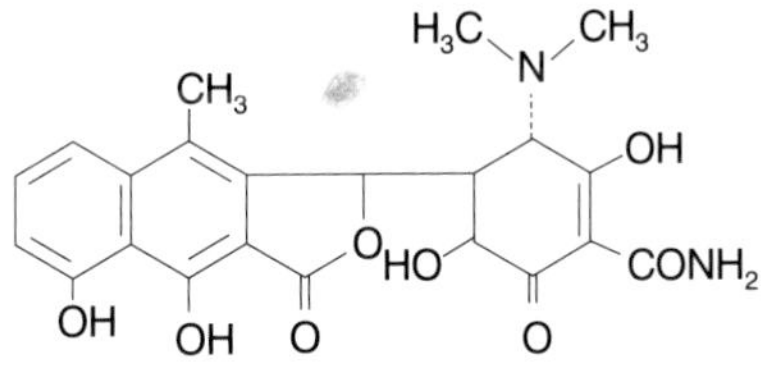

Apo-oxytetracycline

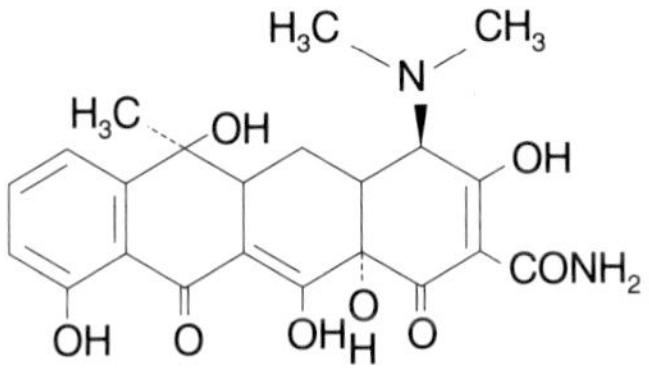

4-Epitetracycline

FIGURE 5.8 Tetracycline derivatives with greatly reduced antibiotic activity.

all been implicated as possible complexing sites for polyvalent cations. An alternative suggestion based on circular dichroism studies on the 7-chlortetracycline complexes with Ca^{2+} and Mg^{2+} is that chelation requires the bending of ring A back towards ring B and C so that the oxygen atoms at positions 11 and 12 together with those at positions 2 (amidic oxygen) and 3 form a co-ordination site into which the metal atom fits. The structural modifications to the tetracycline molecule described above, which affect its ability to inhibit protein synthesis on the ribosomes, may also affect its metal-complexing properties, but no final conclusion as to the nature of the molecular interaction between the tetracyclines and ribosome can be drawn at present.

5.6 Inhibitors of peptide bond formation and translocation

5.6.1 Chloramphenicol

Chloramphenicol (Figure 5.9), a naturally occurring antibiotic that is now produced by total chemical synthesis, has had a chequered history. It was one of the first broad-spectrum antibiotics to be discovered and has excellent bacteriostatic activity against both Gram-positive and Gram-negative cocci and bacilli, together with activity against rickettsias, mycoplasmas and *Chlamydia*. Unfortunately, its ever-widening use in some parts of the world without effective medical controls, revealed serious side-effects associated with the bone marrow in a small number of patients. A concentration of 25–30 μg of chloramphenicol ml^{-1} of blood maintained for 1–2 weeks leads to an accumulation of nucleated erythrocytes in the marrow, indicating an interference with the normal red cell maturation process. Normal erythropoiesis usually resumes after withdrawal of the drug, but very occasionally, i.e. not more than 1 in 20 000 cases, a more serious defect develops in the marrow which leads irreversibly to the loss of both white and red cell precursors. The biochemical basis for chloramphenicol-induced fatal aplastic anaemia has not been established. However, an action of the drug on mitochondrial ribosomes, which more closely resemble 70S than 80S ribosomes, with possible effects on the mitochondrial function of key stem cells in the marrow, cannot be ruled out. The very low incidence of the irreversible form of chloramphenicol toxicity indicates a special sensitivity in those few individuals who succumb to it. Chloramphenicol therapy is therefore now restricted to serious infections for which there may be no effective alternative. These include:

1. typhoid fever and other serious *Salmonella* infections;
2. meningitis caused by bacteria resistant to β-lactam antibiotics or in cases where the patient is allergic to these drugs.

The bacteriostatic action of chloramphenicol is due to a specific inhibition of protein biosynthesis on 70S ribosomes; it is completely inactive against 80S ribosomes. Studies with radioactively labelled chloramphenicol show that it binds exclusively to the 50S subunit to a maximum extent of one molecule per subunit. The binding is completely

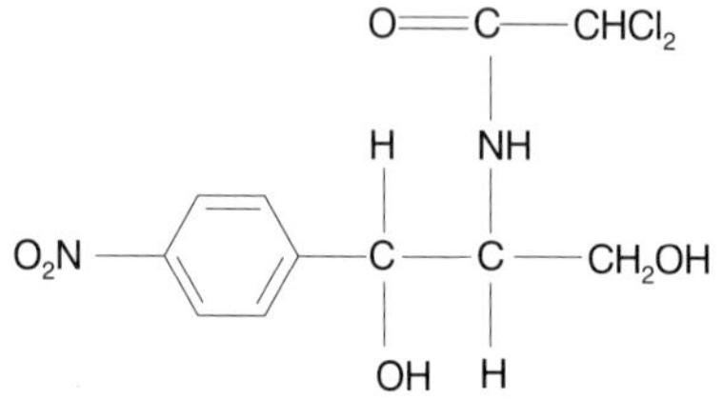

O=C—$CHCl_2$
H NH
O_2N— C — C —CH_2OH
OH H

FIGURE 5.9 Chloramphenicol. The active form is the D-*threo* stereoisomer.

reversible. Structurally unrelated antibiotics such as erythromycin and lincomycin, that also interfere with the function of the 50S subunit, compete with chloramphenicol for the binding region, whereas aminoglycosides and tetracyclines do not. The following biochemical evidence strongly suggests that the primary action of chloramphenicol is to block peptide bond formation by inhibiting the peptidyl transferase activity of the 50S subunit:

1. Chloramphenicol inhibits the puromycin-dependent release of nascent peptides from 70S ribosomes.
2. Chloramphenicol inhibits the puromycin fragment reaction, referred to above, that is catalysed by 50S ribosomal subunits.

Earlier studies concentrated on the identification of proteins thought to be involved in the binding of chloramphenicol to the 50S subunit. The technique of affinity immune electron microscopy suggested that the peptidyl transferase region of the ribosome, containing the proteins L15, L18 and L27, contributed to the binding of chloramphenicol. Because the vast majority of known enzymes are proteins, it is understandable that the early work on the action of chloramphenicol sought to identify a protein or proteins associated with peptidyl transferase activity that would provide the target site for the antibiotic. However, the realization that domain V of 23S rRNA probably catalyses peptide bond formation led to a reassessment of the nature of the target for chloramphenicol. Footprinting studies show that the drug protects several bases located in the central loop region of domain V from alkylation by dimethyl sulphate. These include A-2058, A-2059, A-2062 and U-2506. But how can a molecule as small as chloramphenicol afford protection to several nucleotides in addition to an apparent association with several proteins located near the peptidyl transferase centre? One suggestion is that the binding of the drug to ribosome is a dynamic process which triggers a series of conformational changes resulting in the protection of several nucleotides from alkylation. The process may also permit the interaction of successive chloramphenicol molecules with the various proteins identified by affinity immune electron microscopy. It must be emphasized, however, that this remains speculative at present and the molecular details of the binding of chloramphenicol to 50S subunits have yet to be defined.

5.6.2 The macrolide–lincomycin–streptogramin (MLS) group of antibiotics

This diverse collection of complex, naturally occurring antibacterial antibiotics is often referred to as the MLS group because of similar structural features among some of its members and a pattern of overlapping bacterial resistance to various members of the group. The MLS group also has a shared ability to inhibit protein biosynthesis specifically on 70S ribosomes. The actions of the more important examples of MLS antibiotics will be described individually.

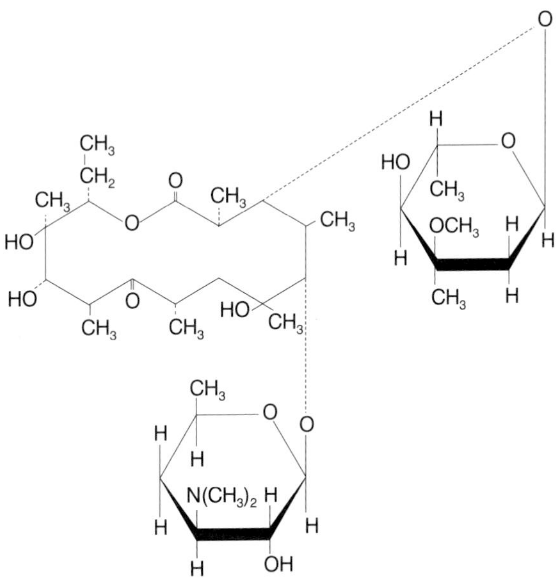

FIGURE 5.10 Erythromycin, a 'medium-spectrum' antibacterial drug of the macrolide family.

Erythromycin

This complex antibiotic (Figure 5.10) is one of many macrolides that are characterized by molecular structures containing large lactone rings linked with amino sugars by glycosidic bonds. Erythromycin is a medium-spectrum antibacterial agent, effective against many Gram-positive bacteria, mycoplasmas and *Chlamydia* but against relatively few Gram-negative organisms. The specific antibacterial action of erythromycin depends on its blockade of protein synthesis on 70S ribosomes while having no action on 80S ribosomes. Like chloramphenicol, erythromycin binds exclusively to the 50S subunit.

Uniquely amongst the antibiotics so far discussed, erythromycin binds specifically to the isolated 50S subunit protein, L15. However, mutations affecting this protein do not cause 70S ribosomes to become resistant to erythromycin, whereas the 50S subunit protein, L22, is altered in *Escherichia coli* mutants selected for erythromycin resistance. While these results indicate a possible contribution of ribosomal proteins to erythromycin binding, evidence from a study of a more commonly encountered form of bacterial resistance to erythromycin strongly suggests the essential involvement of 23S rRNA in the binding event. This type of resistance depends on *N*-dimethylation by the inducible enzyme *N*-methyl transferase of the following equivalent specific adenine residues of 23S rRNA: A-2058 in *Escherichia coli*, A-2086 in *Bacillus stearothermophilus* and A-2058 in *Bacillus subtilis* (see Chapter 9 for further details of the *N*-methyl transferase system). 70S ribosomes reconstituted with *N*-methylated 23S rRNA from resistant cells and ribosomal proteins from susceptible cells are resistant to erythromycin, thus providing a powerful argument for the key contribution of a specific region of 23S rRNA to the target site for the antibiotic. However, molecular details of the putative erythromycin–23S rRNA interaction are unknown at present. Early experiments with the puromycin fragment reaction showed that erythromycin does not directly inhibit peptide bond synthesis but blocks the translocation of peptidyl-tRNA from the A site to the P site. Recent investigations on the mechanism of induction of *N*-methyl transferase by erythromycin suggest that antibiotic bound to or near A-2058 (in *Escherichia coli* ribosomes) blocks the emergence of nascent peptide chains through the exit channel in the ribosome, leading to the destabilization of the ribosome–peptidyl-tRNA complex. Conceivably this destabilization occurs during the attempted translocation of peptidyl-tRNA from the A to the P site, although further research will be needed to confirm this.

Lincomycin and clindamycin

Lincomycin and its clinically more useful analogue clindamycin (Figure 5.11) have a similar antibacterial spectrum to erythromycin. However, while clindamycin has relatively poor activity against mycoplasmas, it is much more effective than erythromycin against anaerobic pathogens such as *Bacteroides* spp.

Both antibiotics bind exclusively to the 50S subunit of the 70S ribosome and have no effect on 80S ribosomes. *N*-dimethylation of A-2058 of 23S rRNA confers ribosomal resistance to lincomycin and clindamycin, thus suggesting that their target sites and mode of action resemble those of erythromycin.

Streptogramins

Antibiotics known as streptogramins were discovered more than four decades ago but until recently were of more academic than practical interest.

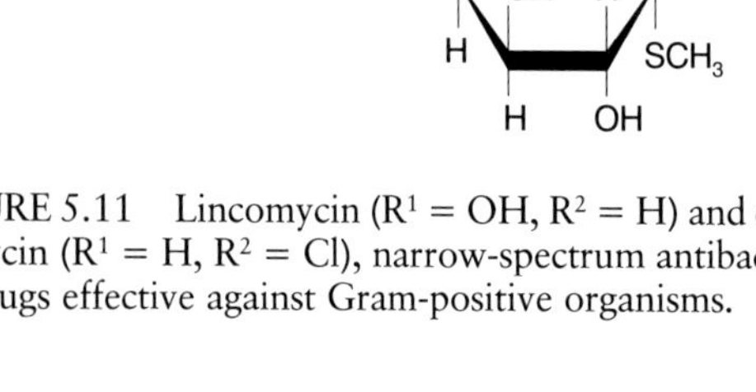

FIGURE 5.11 Lincomycin (R^1 = OH, R^2 = H) and clindamycin (R^1 = H, R^2 = Cl), narrow-spectrum antibacterial drugs effective against Gram-positive organisms.

The growing menace of bacterial resistance to antibiotics has, however, raised the possibility that some members of this large group of complex compounds could be clinically useful. Streptogramins comprise two structurally distinct types which are produced as mixtures by certain *Streptomyces*. Type A streptogramins are polyunsaturated cyclic lactones that resemble the macrolides. Type B streptogramins are cyclic hexadepsipeptides. An example of each type, virginiamycin M_1(type A) and virginiamycin S_1 (type B), is illustrated in Figure 5.12.

Remarkably, the two types of antibiotic act synergistically to give irreversible inhibition of protein biosynthesis and consequently a bactericidal effect. The proposed clinical use of streptogramins would exploit this valuable synergism by combining a type A compound (quinapristin) with a type B agent (dalfopristin). The synergism arises from the distinct actions of the different streptogramins on ribosomal function. Type A compounds block the binding of aminoacyl-tRNA and peptidyl-tRNA to the A and P sites, respectively. Type B compounds hinder only the interaction of peptidyl-tRNA with the P site. Some form of conformational change in ribosomal structure is induced by type A compounds that enhances the affinity of the 50S subunit for type B streptogramins. At present the identity of the binding site for type A compounds has not been established. The situation is somewhat clearer for type B compounds. *N*-dimethylation of A-2058 in 23S rRNA or its replacement by guanine or uridine leads to ribosomal resistance to the latter compounds. It is remarkable that modification or replacement of A-2058 causes resistance to such structurally diverse compounds as erythromycin, lincomycin/clindamycin and type B streptogramins. There is much more to be learned about the precise details of the interactions of all these antibiotics with bacterial ribosomes.

5.6.3 Fusidic acid

Fusidic acid belongs to a group of steroidal antibiotics (Figure 5.13). It inhibits the growth of Gram-

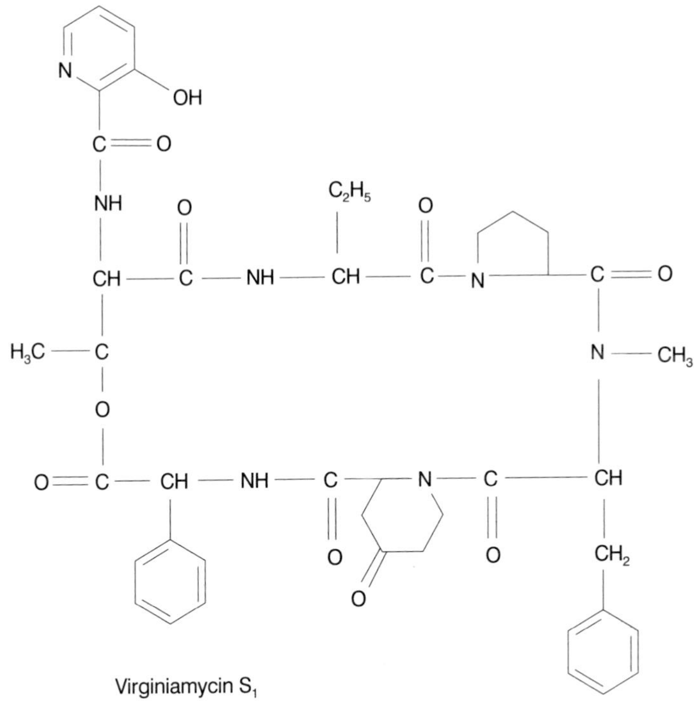

Virginiamycin S_1

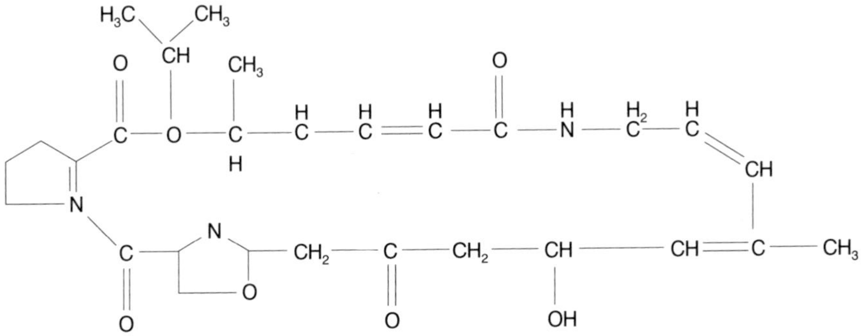

Virginiamycin M_1

FIGURE 5.12 Streptogramin antibiotics that, in combination, cause irreversible inhibition of bacterial protein biosynthesis.

positive but not of Gram-negative bacteria and is occasionally used clinically to treat Gram-positive infections resistant to more widely used agents. The inactivity of fusidic acid against Gram-negative bacteria may be due to inadequate penetration into the bacterial cytoplasm, since the drug inhibits protein synthesis on isolated ribosomes from Gram-negative bacteria.

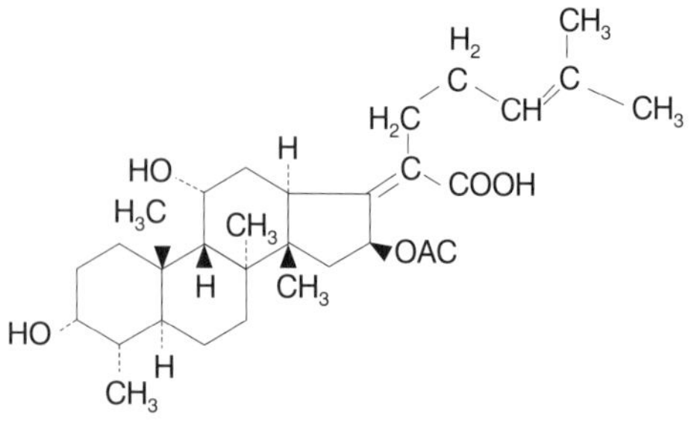

FIGURE 5.13 Fusidic acid, a steroidal antibiotic with a narrow spectrum of action against Gram-positive bacteria.

Addition of fusidic acid to 70S ribosomes *in vitro* prevents the translocation of peptidyl-tRNA from the A site to the P site and also inhibits the EF-G-dependent hydrolysis of GTP. The inhibition is overcome by the addition of excess EF-G. The resistance of some strains of bacteria to fusidic acid appears to be associated with a change in EF-G, since the factor prepared from resistant cells catalyses translocation normally in the presence of the antibiotic. All this points to factor EF-G as the target protein for fusidic acid. However, fusidic acid does not directly inhibit GTP hydrolysis. Indeed, in the presence of the drug there is an initial burst of hydrolysis which then slows to zero. Fusidic acid forms a stable complex with EF-G, GDP and the ribosome, which is unable to release EF-G for a further round of translocation and GTP hydrolysis. Fusidic acid also inhibits protein synthesis on 80S ribosomes in a similar manner by stabilizing the EF-2–GDP–ribosome complex (EF-2 corresponds to the prokaryotic EF-G). The lack of toxicity of fusidic acid against mammalian cells is probably because the drug does not achieve a high enough intracellular concentration in mammalian cells to form the stabilized complex.

5.6.4 Cycloheximide

Occasionally referred to as actidione, cycloheximide (Figure 5.14) is toxic to a wide range of eukaryotic cells, including protozoa, yeasts, fungi and mammalian cells, and its lack of selectivity precludes any clinical use. It is included here because of its unusual specificity of action against 80S ribosomes with no effect on 70S ribosomes. Cycloheximide is therefore mostly used as an experimental tool to inhibit protein synthesis in eukaryotic cells, and occasionally to exclude fungi from bacterial cultures.

There are considerable variations in the sensitivity of 80S ribosomes from different species of yeasts to cycloheximide, which have been exploited to explore its site of action. For example, ribosomes from *Saccharomyces cerevisiae* are strongly inhibited by cycloheximide, while those from *Saccharomyces fragilis* and *Kluyveromyces lactis* are resistant. Cross-over experiments with the 60S and 40S subunits from *Saccharomyces cerevisiae* and *Saccharomyces fragilis* showed that sensitivity to cycloheximide resides in the 60S subunit. Analysis of the proteins of the 60S subunit from *Kluyveromyces lactis* revealed that protein L41 is somewhat different from the corresponding protein in *Saccharomyces cerevisiae,* suggesting that L41 may be involved in the interaction of the antibiotic with susceptible ribosomes. However, it is now believed that the major contributor to the binding site for cycloheximide is the 28S rRNA of the 60S subunit. Footprinting studies show that cycloheximide protects two guanine residues that lie within a loop region associated with the hydrolysis of GTP and ribosomal interaction with elongation factor EF-2. This is consistent with the observed interference of cycloheximide with the translocation of peptidyl-tRNA from the A site to the P site on the larger ribosomal subunit, and it seems likely that the antibiotic hinders the function of EF-2. The role of protein L41 in conferring

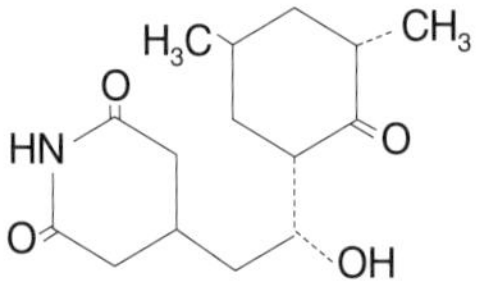

FIGURE 5.14 Cycloheximide, an unusual specific inhibitor of 80S ribosomes.

cycloheximide resistance to ribosomes from *Kluyveromyces lactis* may be due to an ability to modify the conformation of 28S rRNA, thereby protecting a vulnerable domain of the nucleic acid from the inhibitory action of cycloheximide.

5.7 Nitrofurantoin – a ribosomal poison?

Nitrofurantoin (Figure 5.15) is a wholly synthetic antibacterial drug used primarily to treat urinary tract infections. Remarkably, despite 30 years of use there is little or no significant bacterial resistance to nitrofurantoin. The drug is apparently converted by bacterial nitroreductases to a mixture of highly reactive electrophilic derivatives that react with nucleophilic sites on many bacterial macromolecules. Although many aspects of bacterial metabolism are inhibited, perhaps explaining why resistance to the drug has been slow to emerge, there is evidence of a more specific attack on the synthesis of several inducible enzymes, including β-galactosidase, galactokinase and tryptophanase, at minimal antibacterial concentrations.

Recent studies with radiolabelled nitrofurantoin, at concentrations that block inducible enzyme synthesis, revealed that while most ribosomal proteins became radioactive there was preferential labelling of proteins L5, S14 and S18. It is possible that a covalent interaction of S18 with an electrophilic derivative of nitrofurantoin could disrupt the codon–anticodon interactions that take place in the general location of S18 on the platform region of the 30S subunit (Figure 5.1). However, it is difficult to see why this should not cause total inhibition of protein biosynthesis rather than the observed specific suppression of inducible enzyme synthesis. It has been suggested that translation of the mRNA of inducible enzymes may be especially sensitive to secondary structure changes caused by the modification of S18 in the purine-rich initiator (Shine–Dalgarno) region centred about 10 nucleotides on the 5' side of the initiator codon. Further research is needed to establish the validity or otherwise of this hypothesis.

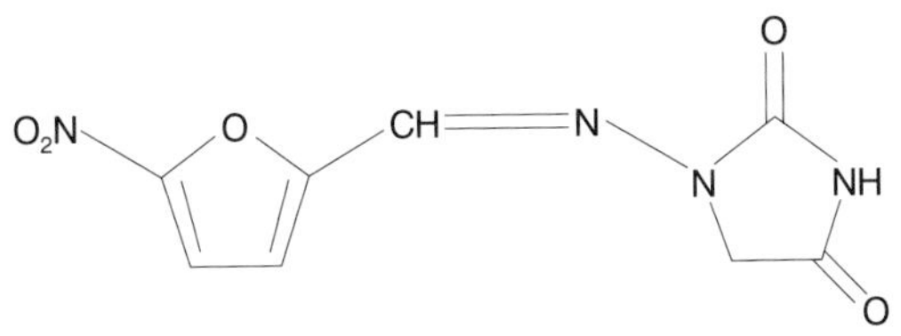

FIGURE 5.15 Nitrofurantoin, a synthetic antibacterial drug used in the treatment of urinary tract infections.

5.8 Effects of inhibitors of 70S ribosomes on eukaryotic cells

Subcellular organelles in eukaryotic cells, such as mitochondria and chloroplasts, contain ribosomes that more closely resemble 70S than 80S ribosomes both in size and in sensitivity to ribosomal inhibitors. Although only a small proportion of the protein of the subcellular organelles is synthesized on the organelle-specific ribosomes, the inhibition of these ribosomes may underlie the action of some antibiotics on eukaryotic cells. For example, yeasts grown in the presence of chloramphenicol, erythromycin or lincomycin become deficient in various cytochromes, although these proteins are not synthesized on mitochondrial ribosomes. Possibly the antibiotics in some way indirectly block the integration of the cytochromes into the mitochondrial structure. Streptomycin and erythromycin cause the loss of the photosynthetic organelles in the algae *Chlamydomonas* and *Euglena* and this may be associated with antibiotic attack on the chloroplast ribosomes. The possibility that the rare but serious toxic side-effect of chloramphenicol on the bone marrow could be associated with an inhibitory action of the drug on the mitochondria of essential stem cells in the marrow has already been discussed.

Further reading

Cannon, M. (1994). Interaction of cycloheximide with 25S rRNA from yeast. *Biochem. Soc. Trans.* **22**, 449S.

Green, R. and Noller, H. F. (1997). Ribosomes and translation. *Ann. Rev. Biochem.* **66**, 679.

Oakes, M. I. *et al.* (1990). Ribosome structure: three dimensional locations of rRNA and proteins. In *The Ribosome: Structure, Function and Evolution* (eds W. E. Hill *et al.*), American Society for Microbiology, Washington, DC, p.180.

Pechére, J.-C., (1996). Streptogramins: a unique class of antibiotics. *Drugs* **51**, Suppl. 1, 13.

Purohit, P. and Stern, S. (1994). Interactions of a small RNA with antibiotic and RNA ligands of the 30S subunit. *Nature* **370**, 659.

Rodriguez-Fonseca, C. (1995). Fine structure of peptidyl transferase on 23S-like rRNAs deduced from chemical probing of antibiotic–ribosome complexes. *J. Molec. Biol.* **247**, 223.

Vannuffel, P. and Cocito, C. (1996). Mechanisms of action of streptogramins and macrolides. *Drugs* **51**, Suppl. 1, 20.

Woodcock, J. *et al.* (1991). Interaction of antibiotics with A- and P-site-specific bases in 16S ribosomal RNA. *EMBO J.* **10**, 3079.

Yanagisawa, T. *et al.* (1994). Relationship of protein structure of isoleucyl-tRNA synthetase with pseudomonic acid resistance of *Escherichia coli*. *J. Biol. Chem.* **269**, 24304.

Chapter six

Antimicrobial drugs with other modes of action

The antimicrobial agents described so far have been arranged together according to their primary effects on cellular metabolism and biosynthesis. This approach accounts for many of the most important drugs in current antimicrobial therapy. There are, however, several valuable drugs whose various modes of action fall outside those most commonly encountered. In this chapter we describe some important examples of compounds with unusual actions which we have arranged according to their therapeutic targets: antibacterial, antifungal, antiviral and antiprotozoal.

6.1 Antibacterial agents

6.1.1 Metronidazole

This synthetic compound (Figure 6.1) is valuable in the treatment of infections caused by strictly anaerobic bacterial pathogens such as *Bacteroides fragilis* and *Clostridium difficile,* and also by *Helicobacter pylori,* a bacterium causally linked to peptic ulcer disease and gastritis and which exists in an environment of low oxygen tension. The spectrum of metronidazole also extends to several protozoal parasites, including *Giardia lamblia, Entamoeba histolytica* and *Trichomonas vaginalis.*

In anaerobic species the compound is absorbed rapidly by the cells and then reduced by cellular metabolism to a short-lived nitro radical anion:

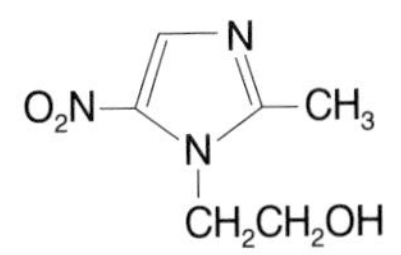

O_2N — N — CH_3 — N — CH_2CH_2OH

FIGURE 6.1 Metronidazole, a synthetic antibacterial useful in the treatment of infections caused by strict anaerobes.

$$\text{R-NO}_2 \rightarrow \text{R-NO}_2^{\bullet -}$$

The radical attacks DNA, causing both single- and double-stranded breaks. Although the precise nature of the damage to DNA is not clear, there is a preferential attack of the reactive metabolite on thymidine residues and other pyrimidines. There is no evidence for the formation of an adduct between the nitro radical anion and DNA. Instead, the metronidazole metabolite is believed to oxidize vulnerable residues in DNA by abstracting electrons. Bacteria with mutations that adversely affect excision repair and DNA recombination are more sensitive to metronidazole. Damage to DNA is believed to be the cause of cell death and lysis in bacteria and protozoal parasites.

The model presented above accounts for the activity of metronidazole against anaerobic bacteria. The situation is less clear in the case of organisms which survive under condition of low, rather than zero, oxygen tension. One suggestion is that in the presence of oxygen the nitro radical anion is converted back to metronidazole and superoxide:

$$R\text{-}NO_2^{\bullet -} \rightarrow R\text{-}NO_2 + O_2^{\bullet -}$$

The superoxide generated in this process is then converted by the enzyme superoxide dismutase to hydrogen peroxide and molecular oxygen. In the presence of a transition metal, either iron or copper, a series of reactions, known collectively as the Haber–Weiss reaction, give rise to the highly reactive hydroxyl radical, $OH^{\bullet}$ which also damages DNA.

Several other structurally related antibacterial and antiprotozoal drugs, including tinidazole and secnidazole, are believed to share a common mode of action with metronidazole.

6.2 A unique antifungal antibiotic – griseofulvin

Fungal infections of the skin, frequently caused by various species of *Trichophyton*, usually respond well to topically applied drugs of the azole class (Chapter 3). However, in cases of azole resistance or where the infection is widespread, systemic treatment with griseofulvin (Figure 6.2) may be necessary. Griseofulvin is an antibiotic produced by *Penicillium griseofulvum* which causes the tips of fungal hyphae to become curled and the suppression of further growth. At the molecular level griseofulvin binds to the intracellular protein tubulin and possibly to the accessory proteins involved in the polymerization of tubulin to form microtubules (the microtubule associated proteins, MAPs). Microtubules participate in the movements of subcellular organelles and in the separation of daughter chromosomes during mitosis in eukaryotic cells. A characteristic property of microtubules is their rapid assembly and disassembly due to the reversible polymerization of the constituent tubulin. This dynamic instability, as it is called, is crucial to the mitotic process. The binding of griseofulvin to tubulin and the MAPs in some way hinders the assembly and disassembly of the microtubules and thereby disrupts cell proliferation.

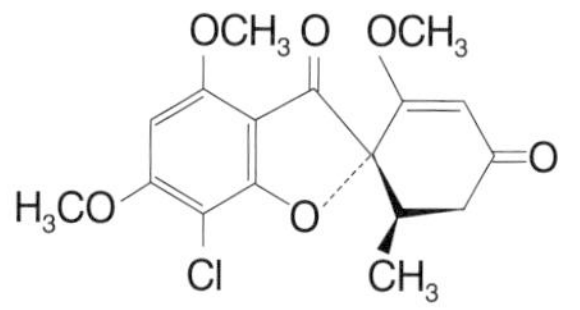

OCH_3 O OCH_3 O H_3CO O Cl CH_3

FIGURE 6.2 Griseofulvin, a unique antifungal antibiotic that disrupts the function of microtubules.

The selectivity of griseofulvin for dermatophytes is not fully understood since the antibiotic also binds to mammalian tubulin. Nevertheless, the many differences in the amino acid sequences of, for example, the tubulin of the yeast *Saccharomyces pombe* and that of mammalian brain tubulin may permit differential binding of griseofulvin to the tubulin of fungal cells under the conditions of antifungal therapy.

6.3 Antiviral agents

6.3.1 Inhibitors of the protease of the human immunodeficiency virus (HIV)

In Chapter 4 we reviewed inhibitors of HIV reverse transcriptase. While these drugs have provided a major advance in the treatment of AIDS, their long-term efficacy in monotherapy is always threatened by the remarkable ability of HIV to generate drug-resistant mutants. To combat this tendency, inhibitors of reverse transcriptase are frequently given in combination with other inhibitors of HIV replication. Outstanding among the latter agents is an expanding group of compounds that inhibit the HIV-specific protease.

The viral translation products formed during the replicative phase of the virus are long polypeptide precursors which are then specifically cleaved to release several mature proteins. In this way the virus generates a number of distinct proteins from a single mRNA molecule. The enzyme responsible for the cleavage process, HIV protease, is itself first formed as a zymogen precursor protein. HIV protease, encoded by the viral *pol* gene, belongs to the aspartyl class of proteases. Enzymes of this mechanistic type are maximally active in the pH range 4.5–6.5 but nevertheless retain significant activity at the cytoplasmic, neutral pH.

As the nature of HIV and its replicative cycle emerged during the 1980s, so the importance of HIV protease as a potential target for antiviral activity became apparent. Considerable attention had already been given to the design of inhibitors of another aspartyl protease, renin, and the peptidic inhibitors of this enzyme provided a good starting point for the development of anti-HIV protease compounds. However, like many of the renin inhibitors, the early HIV protease inhibitors were flawed by extremely low solubility which resulted in poor uptake and tissue distribution ('bioavailability') after dosing. Then in 1989 X-ray crystallography revealed the three-dimensional structure of HIV protease which greatly facilitated the design of non-peptidic inhibitors. Unlike the earlier peptidic inhibitors, several of the non-peptidic compounds proved to have acceptable, although not ideal, bioavailability. As a result, several HIV protease inhibitors are now in clinical use and more compounds are being developed. Both indinavir and ritonavir (Figure 6.3) are potent selective inhibitors that produce rapid falls in the number of viruses in the blood and, significantly, an increase in circulating CD4 lymphocytes that are so critical to effective immune defence against microbial infections. However, it is not yet clear whether the improvement in circulating CD4 cells reflects a similar improvement in CD4 proliferation in lymphoid tissue.

HIV protease inhibitors cannot be given in isolation because of the risk of the rapid emergence of specific viral resistance to these compounds (Chapter 9). Instead they are now established as an essential component of the so-called triple therapy in combination with two different inhibitors of HIV reverse transcriptase. This form of therapy can reduce the circulating viral RNA in 90% of patients to levels that are not detectable by the polymerase chain reaction for up to 1 year. However, it remains to be seen whether the improved status of patients can be maintained indefinitely. Indeed, a recent ominous finding is that replication of the virus may persist in lymph nodes despite sustained drug therapy.

6.3.2 Amantadine: an inhibitor of the initial phase of cell–virus interaction

Amantadine (Figure 6.4) has both prophylactic and therapeutic activity against influenza A_2 infections in mice. In man it offers protection against influenza A_2 infection during the period of dosing, and the duration of the disease appears to be shortened if the drug is started even after the infection has begun. *In vitro*, amantadine inhibits replication of the influenza virus after the virus has entered the host cell but before the process of viral uncoating begins. The mature influenza virus particle is enveloped in a lipid membrane that contains three integral proteins: haemagglutinin, neuraminidase and a protein designated as M_2. This latter protein, which exists as a homo-oligomer of as many as 10–12 M_2 molecules, has been identified as the specific molecular target for amantadine. During the process of viral attack on susceptible cells the virus particles, or virions, enter the cells by the process of endocytosis and are then incorporated into the endosomal compartment. At this stage the M_2 protein is believed to function as a monovalent cation channel across the virion membrane. A proton flux from the endosome through this channel into the virion interior ensures the reduction in internal pH that is required for the uncoating of the virion. The ion channel activity of M_2 has been confirmed with recombinant M_2 protein inserted into the membranes of *Xenopus* oocytes. It now seems fairly clear that the antiviral action of amantadine results from a specific interaction between the drug and the M_2 protein which abrogates its ion-channel function. The precise mechanism of this anti-channel effect is unknown, but it has been suggested that the drug in some way perturbs the conformation of the channel protein oligomer, resulting in a reduction in the frequency with which the channel exists in the open state. Whatever the actual mechanism of the amantadine–M_2 interaction, the end

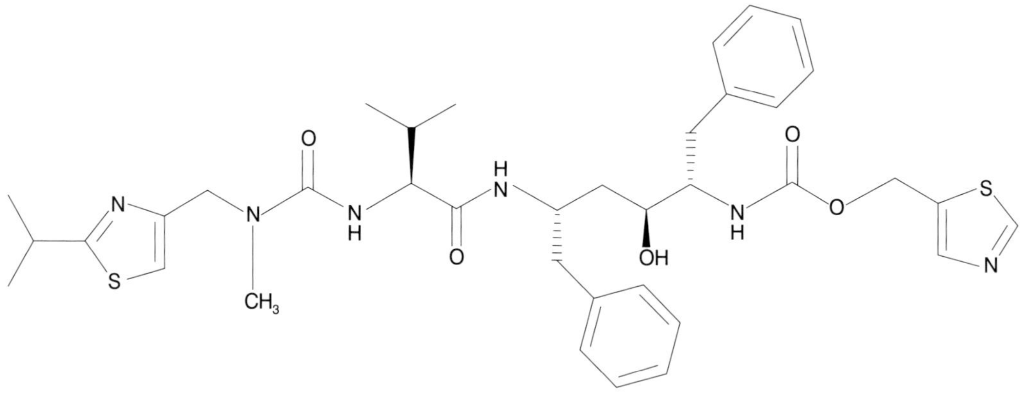

Ritonavir

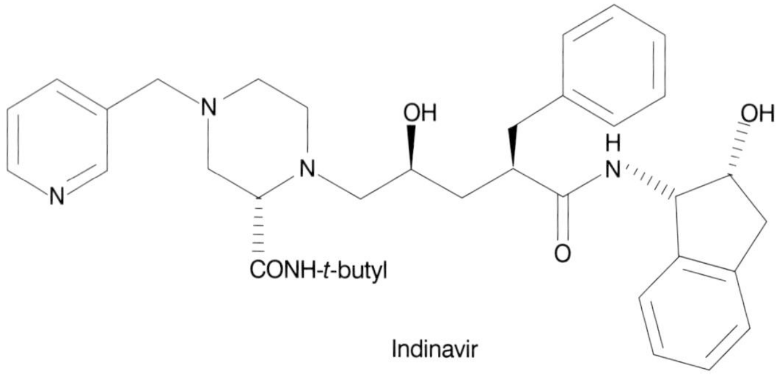

CONH-*t*-butyl

Indinavir

FIGURE 6.3 Inhibitors of HIV protease used in the therapy of AIDS.

result is that the reduction in internal virion pH necessary for virion uncoating is prevented, thus interrupting the cycle of virus replication.

6.3.3 Interferon

Originally discovered in 1957 as a naturally occurring antiviral agent, interferon is now used as the generic name for a family of proteins involved in host defences against certain viral and parasitic protozoal infections. Interferons (IFNs) also affect the immune system, cell proliferation and differentiation and thereby have a useful, though limited, antitumour activity. The therapeutic opportunities for the interferons have expanded considerably with the advent of technology to provide substantial quantities of recombinant proteins.

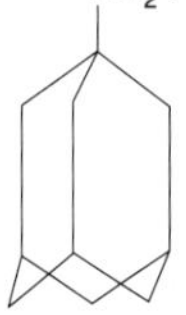

$NH_2 \cdot HCl$

FIGURE 6.4 Amantadine, an anti-influenza drug with a unique action against the ion-channel function of the viral protein M_2.

The most recent nomenclature for IFNs, based on amino acid sequence data, defines four groups: IFNα and IFNω (previously IFNα-1 and IFNα-2), IFNβ and IFNγ. Only IFNα is used therapeutically as an antiviral drug and it is valuable in the treatment of hepatitis B and C. IFNα is also useful against hairy cell leukaemia and AIDS-related Kaposi's sarcoma, actions that may depend upon its antiproliferative rather than its antiviral property. IFNα itself constitutes a family of related proteins encoded by at least 14 functional genes.

The antiviral activity of IFNα is mediated through a complex cascade of events, beginning with the binding of the protein to its specific receptor embedded in the cytoplasmic membrane of the cell. The IFNα receptor consists of at least two subunits, IFNAR1 and IFNAR2. A current model of IFNα receptor function proposes that the R1 and R2 subunits are associated, respectively, with two distinct protein tyrosine kinases, Tyk2 and JAK1 (a 'Janus' kinase). The binding of IFNα to the receptor results in the activation of these kinases which then phosphorylate the gene transcription factors STAT1 and STAT2. The full extent of the transcriptional activity following the activation of STAT1 and STAT2 remains to be elucidated but the major features responsible for the antiviral action of IFNα are reasonably clear.

Three key proteins linked to the antiviral action are induced as a result of the signal transduction process initiated by the binding of IFNα to its receptor:

1. 2′,5′-oligoadenylate synthetase, otherwise known as $(2'\text{-}5')(A_n)$ synthetase, which is active only in the presence of double-stranded (ds) RNA (an intermediate or by-product of viral replication);
2. RNAase L;
3. dsRNA-dependent protein serine kinase.

The $(2'\text{-}5')(A_n)$ synthetase exists in several isoenzymic forms and catalyses the conversion of ATP to a series of AMP oligomers linked by 2′-5′ rather than the usual 3′-5′ phosphodiester bonds. The 2′-5′ A oligomers (up to 15 units in length) then activate the latent form of RNAase L. The active form of this endonuclease hydrolyses both mRNAs and rRNAs at sequences containing UU and UA. The destruction of RNA molecules contributes to both the antiviral and antiproliferative actions of IFNα.

A recent study found that an antisense cDNA directed against the mRNA for $(2'\text{-}5')(A_n)$ synthetase completely suppressed both the antiviral and antiproliferative actions of IFNα, suggesting that activation of the synthetase alone is sufficient for biological activity. However, this study was confined to the action of interferon in various lines of genetically modified mouse fibroblasts infected with mengovirus. A significant antiviral role for the IFNα-induced dsRNA-dependent kinase in other situations cannot therefore be ruled out. The target of this kinase is a specific serine residue in the α subunit of the eukaryotic initiation factor, eIF_2, that mediates the binding of Met-$tRNA_i$ to the 40S ribosomal subunit. Each initiation event requires the hydrolysis of one molecule of GTP bound to eIF_2 to GDP. Phosphorylation of eIF_2 prevents the exchange of the bound GDP for GTP necessary for the next round of initiation.

The inhibitory effects of IFNα on virus growth extend to the various stages of virus infection including penetration, uncoating, transcription, translation and virus assembly. In addition to the three enzymes described above, IFNα induces the expression of many other proteins which may well contribute to the various aspects of antiviral activity. Only further research will complete our understanding of the biological actions of the interferon family.

6.4 Antiprotozoal agents

6.4.1 Antimalarial drugs

Chloroquine

Malaria in its various forms affects some 270 million people each year, mainly in the tropical zones, and as many as 2 million may die of the disease. For more than 50 years chloroquine (Figure 4.10) has been a mainstay of both the prophylaxis and treatment of malaria, although unfortunately resistance to this drug is now widespread (Chapter 9). Early studies suggested that the antimalarial action of chloroquine might depend upon its ability to bind to DNA by intercalation (Chapter 4), leading to inhibition of DNA replication and transcription. However, although it is still possible that the intercalative property of chloroquine may contribute to its antimalarial action, the principal effect of the drug is to disrupt the ability of the parasite to cope with haem released during the metabolism of haemoglobin.

At a certain stage in the complex life cycle of *Plasmodium* parasites, the protozoons invade the red cells of the host. The trophozoites, as the parasitic cells are called at this stage, digest more than 75% of the haemoglobin of the infected red cells in lysosomal vacuoles in order to obtain amino acids for their own development. The digestive process releases the cytolytic porphyrin haem within the parasitic cell. The trophozoites protect themselves against the toxic effect of haem by facilitating its polymerization into the inert, insoluble pigment haemozoin, or β-haematin, in which the iron atom of one molecule of haem is co-ordinated to the propionate carboxyl residue of the next haem. There is considerable debate as to whether this polymerization process is mediated by an enzyme or protein of the trophozoites or whether it may be a 'spontaneous' reaction dependent solely on the chemical nature of haem. However, what is clear is that chloroquine blocks the detoxification of haem by inhibiting its polymerization in some way, so that the trophozoites succumb to the lytic action of haem. The positive charge on the chloroquine molecule is thought to lead to its accumulation within the digestive vacuoles of the parasite to concentrations exceeding 100 μM.

Recent investigations revealed the existence of several histidine-rich proteins (HRPs) in the digestive vacuoles of *Plasmodium falciparum* that promote the polymerization of haem. One of these proteins, HRPII, has 51 His–His–Ala repeats considered to provide the binding sites for the 17 molecules of haem that bind to one molecule of HRPII. Chloroquine inhibits the HRPII-mediated formation of haemozoin. However, against this must be set the finding by other investigators that chloroquine also inhibits an analogous haem polymerization process occurring in the absence of any added proteins. Whether haemozoin synthesis is protein mediated or spontaneous within the trophozoites, it seems likely that the inhibitory effect of chloroquine on the process depends on an interaction between the drug and the haem molecule or minimal haem oligomers.

It had been thought that the more recently introduced antimalarial drug, mefloquine (Figure 6.5), also blocked haemozoin synthesis by a similar mechanism to that of chloroquine. However, it is now believed that the site of action of mefloquine may not be in the digestive vacuoles, since it is not accumulated in the vacuoles in the same way as chloroquine. The precise mechanism of action of mefloquine is unclear but recently detected interactions with specific parasite proteins may be of considerable significance.

Artemisinin and artemether

Artemisinin, extracted from the traditional Chinese herbal medicine qinghaosu, and its close derivative artemether (Figure 6.6), are proving to be useful new weapons in the fight against malaria, particularly against the chloroquine-resistant and cerebral forms of the disease.

The digestion of haemoglobin by the malarial parasite is again critical to the specific antimalarial action of artemisinin and related compounds, although in this case the interaction between artemisinin and the products of haemoglobin digestion results in the formation of cytotoxic, oxidant compounds. The endoperoxide bridge of the drug molecule reacts with the iron atom of haem

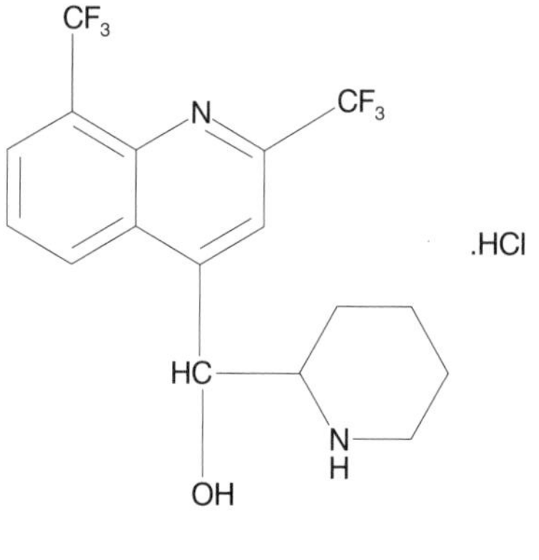

FIGURE 6.5 The antimalarial drug mefloquine.

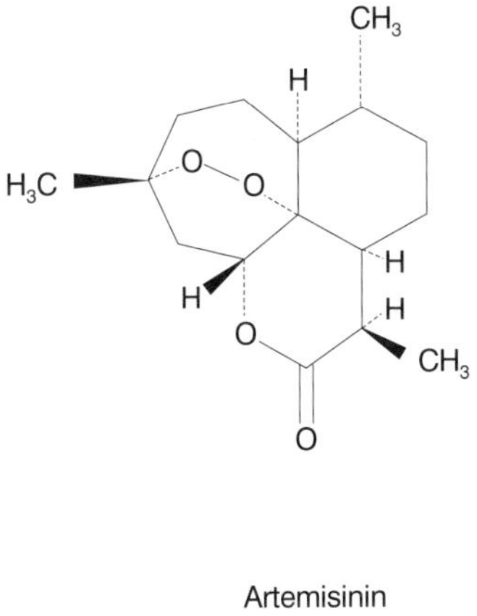

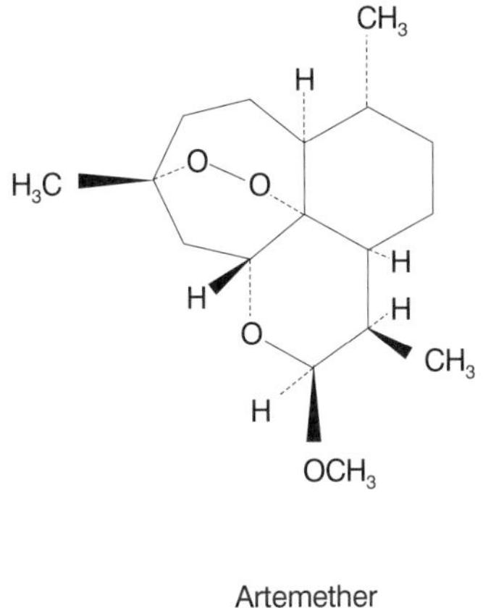

Artemisinin

Artemether

FIGURE 6.6 Two antimalarial agents derived from the traditional Chinese herbal medicine qinghaosu.

released from the digested blood protein. It is possible that artemisinin reacts either with free haem or with the haem residues of the haemozoin polymer. Overall the evidence points to haemozoin as the principal activator of the drug. For example, a strain of *Plasmodium berghei* that lacks haemozoin is highly resistant to artemisinin. Secondly, chloroquine, which blocks the formation of haemozoin by binding to haem, antagonizes that antimalarial action of artemisinin.

The precise details of the interaction between artemisinin and haem are uncertain but the essential role of the endoperoxide bridge is indicated by the fact that derivatives lacking this feature are inactive. Molecular modelling studies suggest that the endoperoxide bridge is in close proximity to the haem iron. The interaction catalyses the breakdown of artemisinin into a free radical that transfers an oxygen atom to the haem iron, generating the oxidant species, $O{=}Fe^{2\bullet}$. This in turn oxidizes lipids and proteins and leads to the death of the parasite. Free radicals also inflict some damage to the red cells, although any toxicity associated with this appears to be minimal and an acceptable price to pay for the therapeutic efficacy of artemisinin against a potentially lethal infection.

Atovaquone

The hydroxynaphthoquinone derivative atovaquone (Figure 6.7) is a relatively recent introduction into the treatment and prophylaxis of malaria and is normally given in combination with an inhibitor of dihydrofolate reductase, such as proguanil. Its unusual mode of action gives atovaquone a useful edge in the struggle to contain chloroquine-resistant malaria. Hydroxynaphthoquinones have long been known for their activity against protozoal parasites, but it was not until the 1990s that the clinical value of atovaquone in malaria was established.

The primary inhibitory action of hydroxynaphthoquinones, including atovaquone, is against the respiratory chain in mitochondria, leading to an interruption in the supply of ATP. Studies with mitochondria prepared from two species of the malarial parasite, *Plasmodium falciparum* and

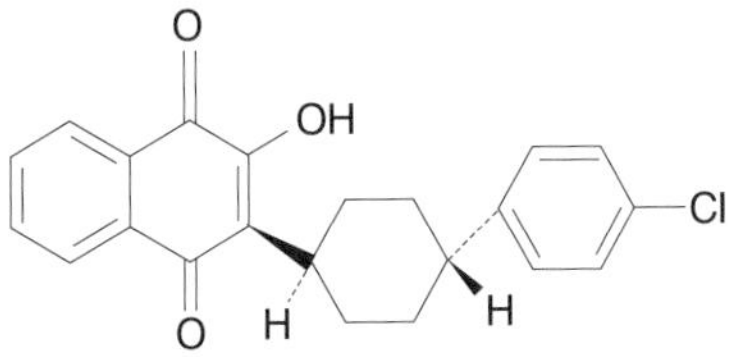

FIGURE 6.7 Atovaquone, a synthetic drug recently introduced into the prophylaxis and therapy of malaria.

Plasmodium yoeli, showed that the cytochrome *c* reductases of α-glycerophosphate (a major substrate for plasmodial mitochondria), succinate and dihydro-oratate were all strongly inhibited by atovaquone. However, the primary site of action in the electron transport chain is on the ubiquinone–cytochrome *c* reductase span of the chain, i.e. cytochrome bc_1, otherwise known as Complex III. There is some evidence that a protein of molecular weight 11.5 kDa associated with Complex III provides a receptor site for atovaquone. A covalent interaction may be involved but the details are unknown. Atovaquone treatment of plasmodia not only depletes the cells of ATP but also depresses their pyrimidine nucleotide content. The latter effect is an indirect result of the inhibition of Complex III.

The antimalarial specificity of atovaquone derives from the much greater sensitivity of the respiratory activity of plasmodial mitochondria to inhibition by the drug compared with that of mammalian mitochondria. The molecular basis of this difference in sensitivity, which may be as much as 1000-fold, is not yet evident

6.4.2 Antitrypanosomal drugs

There are two major forms of trypanosomiasis: African sleeping sickness, caused by two species of *Trypanosoma brucei*, and Chagas disease or South American trypanosomiasis, caused by *Trypanosoma cruzi*. Very few drugs are available to treat these diseases which affect tens of thousands of people in the tropical zones. Trypanosomiasis in Africa is also a major threat to cattle husbandry. The drugs described below, with the exception of eflornithine, were all discovered more than 40 years ago and research into the chemotherapy of trypanosomiasis remains relatively neglected. The action of another antitrypanosomal drug, ethidium, used in veterinary medicine, is discussed in Chapter 4.

Suramin

First made available over 70 years ago, suramin (Figure 1.2) is a large, polysulphonated molecule with six negative charges at physiological pH. It is often the drug of choice for the treatment of the early stages of sleeping sickness. In view of its molecular size and highly charged nature, it is surprising that suramin is able to gain access to the cytoplasm of the parasitic cells. In fact, the compound binds avidly to serum proteins and it is believed that both free proteins and those complexed with suramin enter trypanosomes by the process of endocytosis. The selective toxicity of suramin for trypanosomes may in part be due to a more vigorous uptake of serum proteins by trypanosomes compared with the cells of the infected patient.

Suramin binds to and inhibits a broad range of enzymes derived from *Trypanosoma brucei*, including dihydrofolate reductase, thymidine kinase and all of the enzymes of the glycolytic pathway. Although the IC_{50} values (the concentration of inhibitor needed to inhibit an enzyme by 50%) are in the high range of 10–100 μM, they are much lower than those for the corresponding enzymes from mammalian sources. The glycolytic enzymes of the trypanosome, which generate all of its ATP, are confined to membrane-bounded organelles called glycosomes. Although the physical characteristics of suramin make it unlikely to diffuse into glycosomes, the drug probably binds to the newly synthesized proteins during their cytoplasmic phase prior to entry into the glycosomes. The normal turnover of uncomplexed enzymes in the glycosomes would be expected to lead to their gradual replacement by suramin-bound enzymes entering from the cytoplasm. This model is consistent with the observed progressive slowing down of energy metabolism in cells treated with suramin. The antitrypanosomal action of suramin is unlikely to depend on the inhibition of a single enzyme, a conclusion that is supported by the fact that resistance to suramin has not been a serious problem despite 70 years of use. A multifaceted mode of action probably hinders the ability of trypanosomes to develop resistance to the drug.

Melarsoprol

Otherwise known as melarsen oxide, melarsoprol (Figure 6.8) is a trivalent organic arsenical drug that has been used to treat sleeping sickness since 1949. Unlike suramin, melarsoprol is useful against late-stage disease, although treatment is fraught with the risk of serious side-effects, especially of a potentially lethal encephalopathy.

Although African trypanosomes incubated with melarsoprol die within minutes, the mechanism of action is uncertain and may have more than one aspect to it. Several glycolytic enzymes are inhibited by melarsoprol, including phosphofructokinase (K_i < 1 μM), fructose-2,6-diphosphatase (K_i = 2 μM) and, to a lesser extent, pyruvate kinase (K_i > 100μM). Melarsoprol also disrupts the function of a unique trypanosomal biochemical, trypanothione, [N^1,N^8-*bis*(glutathionyl)-spermidine] with which the drug forms a stable adduct. Trypanothione is a major cofactor in the control of the redox balance between thiols and disulphides in trypanosomes. The adduct with melarsoprol inhibits trypanothione reductase (K_i = 17.2 μM) which is a key enzyme regulating the redox state of trypanothione itself. However, the relatively high K_i casts some doubt on the relevance of the inhibition of trypanothione reductase to the antitrypanosomal action of melarsoprol. Furthermore, incubation of *Trypanosoma brucei* cells with the drug leads to only a minor conversion of reduced trypanothione to its adduct with melarsoprol.

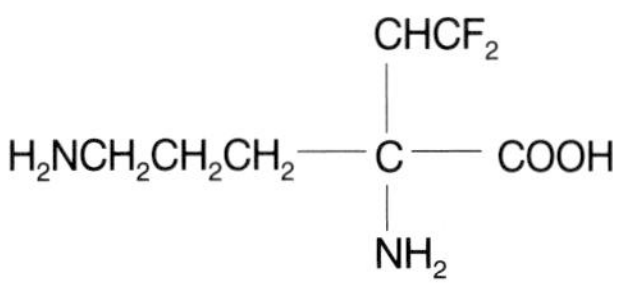

Melarsoprol is typical of organic arsenical agents in its ability to form adducts with thiols and may therefore inhibit many enzymes with essential thiol groups or those that require thiol-containing cofactors. Thus although the major effect of melarsoprol is probably to suppress glycolysis in African trypanosomes, the inhibition of other enzymes may well contribute to the overall antitrypanosomal action. The toxicity of melarsoprol to the patient is also almost certainly due to the avidity of the drug for thiols.

Eflornithine

DL-α-Difluoromethylornithine (DFMO) or eflornithine (Figure 6.8) was designed specifically as a 'suicide' inhibitor of ornithine decarboxylase (ODC) in the hope that the compound would have useful antitumour activity. ODC is involved in the synthetic pathway leading to the polyamines putrescine, spermidine and spermine, which are essential for cell division. Depletion of cellular polyamines caused by the inhibition of ODC results in the suppression of mitosis and cell proliferation. In the event, however, eflornithine turned out to be a better antitrypanosomal drug than an anticancer agent. Somewhat surprisingly, in view of its biochemical action, eflornithine is an extremely safe drug despite having to be given in large doses several times daily for 16 days in order to treat both early and late-stage sleeping sickness.

The basis of the trypanosome-selective action of eflornithine hinges on marked differences in the turnover rates of ODC in mammals and the trypanosome. Eflornithine is an irreversible inhibitor of mouse ODC, forming a covalent adduct with cysteine-360. Since human ODC shares 99% sequence identity with the mouse enzyme it can be reasonably assumed that the human enzyme is also inhibited by

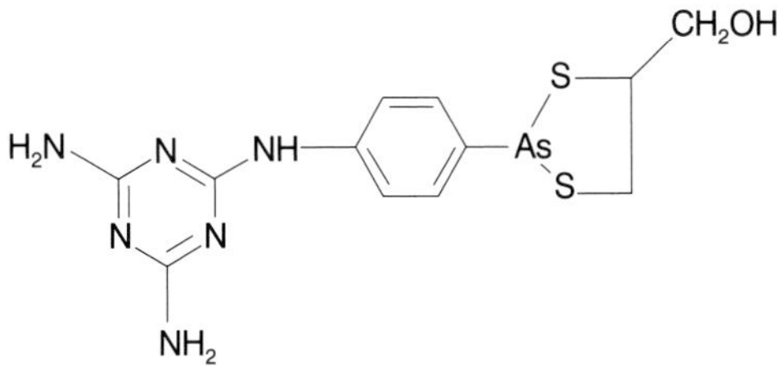

FIGURE 6.8 Two drugs, one old (melarsoprol) and one new (eflornithine), used to treat African trypanosomiasis.

eflornithine in the same way. The K_i against mouse ODC is 39 μM, compared with 220 μM against the corresponding enzyme from *Trypanosoma brucei*. This might lead one to expect that the drug would be more effective against mouse and mammalian cells in general than against trypanosomes. However, while the trypanosomal enzyme is highly stable with minimal intracellular turnover, mammalian ODC has a half-life of only 20 min, placing it among the most rapidly metabolized of eukaryotic proteins. This, combined with the rapid elimination of eflornithine from the body, results in a single dose of the drug exerting only transient inhibition of the constantly renewed ODC of the host. By contrast, there is sustained inhibition of the trypanosomal ODC, resulting in depletion of putrescine and spermidine and also of trypanothione (recall that the latter is a conjugate of glutathione with spermidine). The drug-treated trypanosomes cease dividing and also become incapable of changing their variant surface glycoprotein (VSG). Normally, continual changes in VSG provide the basis of the remarkable ability of trypanosomes to evade immunological detection and destruction by the host. By preventing the changes in VSG the parasite becomes vulnerable to immunological attack and the infection is resolved.

The antitrypanosomal drugs described above are effective only against the African forms of trypanosomiasis. There is little if any useful therapy against Chagas disease. Control of the insect vector for *Trypanosoma cruzi*, the reduviid bug which infests poorly constructed housing, is the preferred approach to the prevention of Chagas disease.

Further reading

Chaudhuri, A. R. and Ludeña, R. F. (1996). Griseofulvin: a novel interaction with brain tubulin. *Biochem. Pharmacol.* **51**, 903.

Cohen, J. L. (1996). Protease inhibitors: a tale of two companies. *Science* **272**, 1882.

Dachs, G. V., Abatt, V. R. and Woods, D. R. (1995). Mode of action of metronidazole and a *Bacteroides fragilis metA* resistance gene in *Escherichia coli*. *J. Antimicrob. Chemother.* **35**, 483.

Egan, T. J., Ross, D. C. and Adams, P. A. (1994). Quinoline antimalarial drugs inhibit spontaneous formation of β-haematin (malaria pigment). *FEBS Letters* **352**, 54.

Fry, M. and Pudney, M. (1992). Site of action of the antimalarial hydroxynaphthoquinone, 2[*trans*-4-(4'-chlorophenyl)cyclohexyl]-3-hydroxy-1,4-naphthoquinone (566C80). *Biochem. Pharmacol.* **43**, 1545.

Jefford, C. W. (1997). Peroxidic antimalarials. *Adv. Drug Res.* **29**, 271.

Pinto, L. H. (1995). Understanding the mechanism of the antiinfluenza drug amantadine. *Trends Microbiol.* **3**, 271.

Sen, G. S. and Lengyell, P. (1992). The interferon system: a bird's eye view of its biochemistry. *J. Biol. Chem.* **267**, 5017.

Wang, C. C. (1995). Molecular mechanisms and therapeutic approaches to the treatment of African trypanosomiasis. *Ann. Rev. Pharmacol. Toxicol.* **35**, 93.

Chapter seven

Penetrating the defences: how antimicrobial drugs reach their targets

In order for a drug to inhibit microbial growth it must reach an inhibitory concentration at its target site. The drug has therefore to penetrate the various permeability barriers that separate its target site from the external environment. Differences in the properties of these permeability barriers among the various species of micro-organisms are important in determining the antimicrobial spectrum of a drug. For example, a specific isolated target site prepared from different bacteria may be inhibited to a similar extent by an antibacterial agent *in vitro*, whereas the intact organisms may exhibit a range of sensitivities to the same drug. This can often be explained by species differences in the structure and composition of the cell envelopes that influence the access of drugs to the target sites. As we shall see, the intracellular concentrations of antimicrobial drugs can also be profoundly affected by the activities of drug efflux systems.

Acquired resistance to some antimicrobial agents depends upon diminished drug accumulation compared with that in wild-type organisms. The change in accumulation can be exquisitely selective, affecting only the concentration of one particular inhibitor in the cell, while in other cases a generalized non-specific decrease in cellular permeability hinders the access of all drugs. The relationship of acquired drug resistance to changes in drug accumulation is described in greater detail in Chapter 8.

7.1 Cellular permeability barriers to drug penetration

7.1.1 The cytoplasmic membrane

The permeability barrier provided by the cytoplasmic membrane is common to all cellular micro-organisms. Its exact composition depends very much on the cell of origin but the most important feature is the lipid bilayer found in all cytoplasmic membranes. Drugs may cross this barrier either by passive diffusion or by facilitated diffusion involving a biological carrier system.

Passive diffusion

The rates of passive diffusion of uncharged organic molecules across lipid membranes are governed by Fick's law of diffusion and correlate reasonably well with their lipid/water partition coefficients. Fick's law is expressed by the equation:

$$V = P \times A(S_e - S_i)$$

where V is the rate of diffusion (in nmol mg^{-1} s^{-1}), A is the surface area of the membrane (cm^2 mg^{-1}), S_e and S_i are respectively the external and internal concentrations of free permeant, and P is the permeability coefficient (cm s^{-1}). It should be noted that the internal concentration of free drug can be substantially affected by binding to intracellular targets, metabolism to other chemical species or by changes in ionization due to differences between internal and external pH values. Lowering of the internal free concentration of a compound by any of these factors enhances the rate of inward passive diffusion.

The higher the lipid solubility of a compound, expressed as the partition coefficient, the more readily it enters and diffuses across the membrane. However, when lipid solubility is so high that a compound is virtually insoluble in water, it may be unable to diffuse from the lipid interior of the membrane into the aqueous environment of the cytoplasm. This adversely affects the biological activity of compounds with sites of action within the cytoplasmic compartment. The relationship between biological activity and lipophilicity is expressed in the 'Hansch equation' (so named after the scientist who formulated it):

$$\log (1/C) = -k(\log P)^2 + k' \log P + \rho\sigma + k''$$

where C is the molar concentration of the drug for a standard biological response, in the case of antimicrobial drugs usually the minimal inhibitory concentration (MIC) or, alternatively, the concentration needed for 50% inhibition of growth or cell survival (IC_{50}); P is the partition coefficient; ρ and σ are physico-chemical constants (Hammett constants) defining certain electronic features of the molecule; and k, k' and k'' are empirically determined constants. The equation indicates that in a chemically related series of biologically active molecules with similar values of ρ and σ, an optimal partition coefficient is associated with maximum biological activity. However, this relationship holds only for those agents that cross membranes by passive diffusion and it may break down when biologically facilitated transport is involved. Furthermore, in some situations the lipophilicity of a drug can be critical in determining binding to its target site if hydrophobic forces dominate the interaction.

The Hansch equation has been applied to various sets of synthetic antibacterial compounds that penetrate the bacterial envelope by passive diffusion. The results show that the most active compounds against Gram-negative bacteria are generally less lipophilic than compounds active against Gram-positive organisms. The cytoplasmic membranes of the two classes of bacteria are sufficiently similar to make it unlikely that they could account for the differences in the partition coefficients of optimally active compounds. A more likely explanation is to be found in the unique properties of the of the outer membrane of Gram-negative bacteria (Chapter 2). We shall return to this important topic later in the chapter.

The rates of passive diffusion of water-soluble molecules across lipid membranes are usually very low, although compounds with molecular weights of less than 100 Da move across as though the membranes were interspersed with water-filled channels or pores. These notional water-filled channels of the cytoplasmic membrane are distinct from the larger, well-characterized hydrophilic pores formed by the porin proteins in the outer membrane of Gram-negative bacteria (Chapter 2). Although water-soluble molecules of molecular weight up to 600 Da or so diffuse through the pores of the outer membrane, the molecular size of most antibacterial agents prevents them from entering the water-filled channels of the cytoplasmic membrane. Nevertheless, as we shall see, certain hydrophilic antibiotics readily enter the bacterial cytoplasm.

Ionized compounds with molecular weights greater than 100 Da diffuse across cytoplasmic membranes with difficulty unless there are compensatory lipophilic regions within the molecules.

This is because the ionized groups in aqueous solution possess strongly bound hydration shells that hinder diffusion across the lipid bilayer. The effect of ionization on the activity of an antibacterial agent is well illustrated by erythromycin (Figure 5.10). The pK_a of the basic dimethylamino group of this antibiotic is 8.8 and the concentration required for antibacterial activity decreases markedly as the pH of the bacterial medium is increased from neutrality towards 8.8. Very likely only the unionized form of erythromycin penetrates into the bacteria and this represents an increasing proportion of the total erythromycin as the pK_a of the drug is approached.

Facilitated diffusion

A remarkable feature of cytoplasmic membranes is their ability to transfer certain physiologically important molecules at much higher rates than could occur by passive diffusion. This process, which is especially notable with biologically essential water-soluble and ionized molecules, is known as facilitated transfer or facilitated diffusion. Characteristically, the rate of transfer of the permeant is proportional to its concentration over a limited range beyond which a limiting rate is approached. This is due to the involvement in the transfer process of carrier proteins within the membrane that are permeant-specific. The rate of transfer increases with increasing permeant concentration until all of the carrier sites are saturated. This is in contrast with passive diffusion where the transfer rate is proportional to permeant concentration over a much wider range. Facilitated transfer by itself results in the equilibration of the permeant across the membrane. However, when the transfer system is linked to an input of 'energy', usually the hydrolysis of ATP or the proton motive force across the cytoplasmic membrane, the permeant is transferred across the membrane against its concentration gradient. This is known as 'active transport'. Facilitated transfer systems are often highly specific and only close structural analogues of the natural permeant compete effectively for the transport sites. These transport systems are vitally important for the uptake of nutrients from the extracellular environment. From this we would expect facilitated transfer and active transport to be used only by drugs resembling natural cellular nutrients, and this is generally borne out in practice.

7.1.2 The outer layers of bacterial cells

Although the cytoplasmic membrane is an important barrier to the penetration of many antimicrobial agents, it provides only a partial basis for the differences in the antibacterial spectra of some antibiotics. Table 7.1 lists a number of agents that are active against both wild-type Gram-positive and Gram-negative bacteria and also several that are usefully active only against Gram-positive organisms. The poor activity of this latter group of drugs against Gram-negative bacteria is due partly to slow penetration of the complex outer layers of Gram-negative cells and also to the existence of drug efflux systems that remove drugs from the cytoplasm of some Gram-negative bacteria (see below).

The structures of the cell envelopes of the two main bacterial groups are described in Chapter 2. In neither group is the peptidoglycan layer a significant obstacle to drug entry. Because of their strongly polar, predominantly negatively charged

TABLE 7.1 Differential sensitivity to typical antibacterial drugs

Drugs active against Gram-positives and Gram-negatives	*Drugs less active against Gram-negatives*
Tetracyclines	Benzylpenicillin
Streptomycin and aminoglycosides	Methicillin
Albomycin (a sideromycin)	Macrolides
Sulphonamides	Lincomycin
D-Cycloserine	Rifamycins
Chloramphenicol	Fusidic acid
Fosfomycin	Vancomycin
Many synthetic antiseptics	Novobiocin
Nitrofurans	Bacitracin
Ampicillin and carbenicillin	
Thienamycin	

nature, the teichoic acids of Gram-positive cells could, in principle, influence the penetration of ionized molecules. The interaction of water-soluble, positively charged compounds, such as the aminoglycosides, with teichoic acid might generate locally high drug concentrations within the envelope, enabling the drugs to challenge the permeability barrier of the cytoplasmic membrane more effectively. In contrast, the entry of anionic molecules could be retarded by teichoic acid, although the exquisite sensitivity of many wild-type Gram-positive bacteria to penicillins, which are organic anions, shows that the repulsive effect of teichoic acid is not very significant.

In contrast, the outer membrane of Gram-negative bacteria presents a significant permeability barrier to many antibacterial agents. The first indication that this membrane hinders drug penetration came from studies with Gram-negative cells with defective envelopes. L-Phase (or L-forms) of *Proteus mirabilis* were found to be 100 to 1000 times more sensitive than intact cells to erythromycin and several other macrolides. There was a smaller increase in sensitivity to other antibiotics, including streptomycin, chloramphenicol and the tetracyclines. Both the outer membrane and the peptidoglycan are also defective in L-forms so that the relative contributions of the various outer layers of intact bacteria to the barrier function was uncertain in these early studies. However, many other studies since have clearly defined the outer membrane as the important permeability barrier.

The structure of the outer membrane

Two major features of the lipid components of the outer membrane of Gram-negative bacteria distinguish it from the cytoplasmic membrane:

1. Negatively charged lipopolysaccharide (LPS) in the outer leaflet of the bilayer replaces the glycerophospholipid of most other biological membranes. The negative charge of the LPS is partly neutralized by divalent cations, mainly Mg^{2+} and Ca^{2+}, which are readily removed by chelating agents such as ethylenediaminotetraacetic acid (EDTA).
2. All the fatty acids of the LPS are unsaturated and this probably makes the interior of the membrane considerably more rigid than that of the cytoplasmic membrane, where there is a much higher proportion of saturated fatty acids. An additional factor restricting the fluidity of the outer membrane is that LPS contains six or seven covalently linked fatty acid chains whereas glycerophospholipid has only two fatty acid residues linked to the head group.

These unique features of the outer membrane slow the diffusion of hydrophobic (i.e. lipophilic) molecules across the lipid bilayer. Many antibacterial agents have hydrophobic domains which hinder their penetration of the outer membranes of Gram-negative bacteria.

Movement of hydrophilic molecules across the outer membranes of Gram-negative bacteria is largely through water-filled pores penetrating the lipid bilayer. These are formed by the porin proteins, the natural function of which is to facilitate the uptake of water-soluble nutrients. Porin proteins consist of trimers of β-barrels arranged as antiparallel β-strands threading through the outer membrane. At the inner face of the membrane the strands are joined by short β-turns and at the outer face by longer loops of amino acids. Three types of porin channel have been identified:

1. general channels with low permeant selectivity;
2. permeant-selective channels with internal specific binding sites;
3. permeant-selective 'gated' channels that only open upon the binding of the specific permeant.

All three types of porin channel restrict transit to compounds with molecular weights of less than approximately 600 Da. Although water-soluble antibiotics usually pass through the general channels, the selective porins are also used. The β-lactam antibiotic imipenem, for example, diffuses through the basic amino acid-specific channel OprD in *Pseudomonas aeruginosa*. The movement through the general porin channels of β-lactam antibiotics approaching the molecular weight limit is retarded by repulsive interactions between the drugs and the predominantly negatively charged amino acids lining the channels. These interactions may hinder diffusion by as much as 100-fold. The importance of porin channels to the influx of hydrophilic antibacterial agents is demonstrated clearly by the reduced susceptibility of porin-deficient mutants to antibiotics, including some valuable semi-synthetic β-lactams. The diffusion of lipophilic molecules through the porin channels is much more difficult because the charged amino acid residues lining the narrowest regions of the channels orient their associated water molecules in a direction that hinders the passage of lipophilic permeants.

Several species of Gram-negative bacteria, most notably the potentially dangerous opportunist pathogen *Pseudomonas aeruginosa*, are deficient in the porin pathway. The high-flux channels of *Escherichia coli* are replaced in *Pseudomonas aeruginosa* by a low-efficiency porin that restricts diffusion to about 1% of the rate through the channels of other Gram-negative bacteria. The absence of efficient porin channels and the low permeability of the rigid LPS of its outer membrane are major contributors to the characteristic intrinsic resistance of *Pseudomonas aeruginosa* to both hydrophilic and lipophilic agents.

It had been thought that the barrier function of the outer membrane provided an adequate explanation for the intrinsic resistance of Gram-negative bacteria to many drugs. However, it is now recognized that even a highly effective permeability barrier cannot completely stem the influx of drugs. An interesting example is provided by hydrophilic β-lactams. These compounds cross the outer membrane of *Escherichia coli* through the porin system and also by slow diffusion across the lipid bilayer. The half-equilibration time of these compounds into the periplasmic space is less than 1 s. In porin-deficient mutants the only route of drug ingress is by diffusion across the lipid bilayer. Even in this situation the half-equilibration time is only a few minutes. Thus although the minimal inhibitory concentrations of β-lactams against the porin-deficient cells are significantly increased, there is still effective access to the penicillin-binding proteins on the outer face of the cytoplasmic membrane.

7.2 Multidrug efflux in bacteria

Even the very low permeability of the outer membrane of *Pseudomonas aeruginosa* cannot account completely for its exceptional resistance to many antibacterial agents. There must therefore be some additional factor at work. In fact, wild-type strains of this organism can extrude drugs as different as tetracycline, chloramphenicol and norfloxacin from the cytoplasm into the external medium by various drug efflux systems. The efflux activities of the various strains correlate well with their individual levels of antibiotic resistance, although the absolute levels of resistance are determined by synergism between the low permeability of the outer membrane and the drug efflux systems. The operon *mexA–mexB–OprM* encodes a broad specificity transporter in *Pseudomonas aeruginosa* capable of extruding a wide range of compounds. The proteins MexB, MexA and OprM, embedded respectively in the cytoplasmic membrane, periplasmic space and outer membrane, act in a co-ordinated manner to remove drugs from the cytoplasm into the external environment. Figure 7.1 illustrates how this system may be arranged in the cell envelope. Although the actual mechanism of drug extrusion is unknown at present, it is believed that the MexA protein provides a link between MexB, a proton antiporter containing 12 trans-cytoplasmic α-helices, and the OprM protein, which may form a channel in the outer membrane. The outward transport of drugs against their concentration gradients is energized by the proton motive force across the cytoplasmic membrane. Mutational inactivation of either MexA or OprM results in a marked increase in cellular sensitivity to a wide range of antibacterial agents. In contrast, overexpression of *mexAB–OprM* in *Pseudomonas aeruginosa* is responsible for the widespread occurrence of carbenicillin-resistant clinical isolates. Since carbenicillin cannot diffuse across the cytoplasmic membrane due to its highly charged nature (it carries two ionized carboxyl groups), the MexAB components of the efflux pump presumably remove carbenicillin from the periplasmic space. The regulation of drug efflux systems involved in acquired drug resistance is considered in greater detail in Chapter 8. It is important to realize, however, that the effectiveness of the contribution of the multidrug efflux system to the intrinsic drug resistance of wild-type *Pseudomonas aeruginosa* depends on the slowness of the inward movement of drugs across the permeability barrier of the outer membrane. When this barrier function is breached by mutational change or by treatment with the chelating agent EDTA, which chelates divalent cations and thereby disrupts the LPS of the outer membrane, the efflux system is overwhelmed by the inward rush of drug molecules and the high level of intrinsic resistance is lost.

Multidrug efflux systems are not confined to *Pseudomonas aeruginosa.* A somewhat analogous

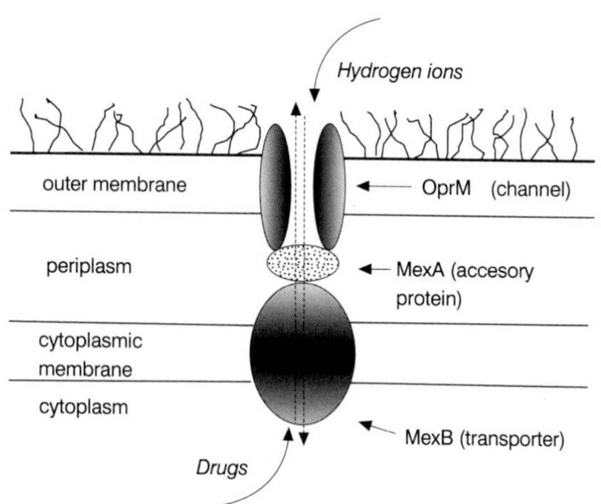

FIGURE 7.1 Diagrammatic representation of the possible arrangement of the components of the multi drug efflux system in Gram-negative bacteria encoded by the *mexA–mexB–OprM* operon. The MexB transporter is a protein of 12 trans-cytoplasmic-membrane domains. The MexA protein acts in some way to provide a link between transporter protein and the OprM channel in the outer membrane. The components are not drawn to scale.

system, designated AcrAB, is found in wild-type *Escherichia coli.* In contrast with the MexAB–OprM system, however, the AcrAB complex only causes intrinsic resistance to large lipophilic drugs, such as erythromycin, fusidic acid and detergents, that traverse the porin channels with difficulty. Susceptibility to smaller antibiotics, including tetracyclines, chloramphenicol and fluoroquinolones, that diffuse rapidly through the porin channels, remains high. The rate of influx of these drugs overwhelms the capacity of the AcrAB system to maintain cytoplasmic concentrations below inhibitory levels. Nevertheless, deleterious mutations affecting AcrAB greatly enhance the sensitivity of *Escherichia coli* to a variety of drugs.

The *acrA* and *acrB* genes of *Escherichia coli* form an operon that encodes AcrA, a periplasmic protein, and AcrB, another 12 transmembrane α-helix transporter embedded in the cytoplasmic membrane. Unlike its counterpart in *Pseudomonas aeruginosa,* the *acrAB* operon does not contain a gene for an outer membrane channel protein. However, the ability of the AcrAB system to pump drugs directly into the external medium suggests that an outer membrane channel may be associated with AcrAB. Homologues of both *acrAB* and *mexAB–OprM* are widespread amongst Gram-negative bacteria.

Multidrug efflux systems are also found in Gram-positive bacteria, for example the QacA system (a 14-transmembrane-domain protein) in *Staphylococcus aureus* which extrudes antiseptic compounds such as chlorhexidine and cetyltrimethylammonium bromide, and the NorA, 12-transmembrane-domain protein, also found in *Staphylococcus aureus,* which includes the fluoroquinolones and chloramphenicol in addition to antiseptics in its range of substrates. However, multidrug efflux in Gram-positive bacteria is probably more relevant to acquired rather than to intrinsic drug resistance. Low-level expression of efflux systems in wild-type Gram-positive cells would be relatively ineffective in conferring intrinsic resistance because of the lack of an outer permeability barrier.

7.3 Examples of the uptake of antibacterial drugs

The key features of bacterial envelopes that influence the uptake of antibacterial agents are summarized in Table 7.2. As we have seen, the inward diffusion of antibacterial agents across the outer membrane of Gram-negative bacteria is essentially a passive process. Some lipophilic drugs, including sulphonamides, rifamcyins and macrolides, probably also diffuse passively across the cytoplasmic membranes of both Gram-negative and Gram-positive bacteria. Hydrophilic compounds, however, are unlikely to achieve inhibitory intracellular concentrations by passive diffusion across the intact cytoplasmic membrane unless there is some form of intracellular sequestration to prevent back diffusion across the cytoplasmic membrane. A few drugs resembling natural nutrients are subject to facilitated diffusion and/or active transport. Examples of the various modes of transport into the bacterial cytoplasm are described below.

7.3.1 D-Cycloserine

This antibiotic, a structural analogue of D-alanine (Figure 2.10), is transported across bacterial cytoplasmic membranes by alanine permeases. In *Streptococcus faecalis* D- and L-alanine use the same transport system and D-cycloserine competitively inhibits the uptake of both isomers. In *Escherichia coli*, where D- and L-alanine are transported by separate isostere-specific permeases, D-cycloserine exploits the D-isomer-specific system. There are also high- and low-affinity transport systems for D-alanine, and at minimal antibacterial concentrations (~4 μM) D-cycloserine favours the high-affinity carrier. The high-affinity D-alanine permease is energy-coupled to the proton motive force, causing accumulation of both D-alanine and D-cycloserine against their concentration gradients. The accumulated D-cycloserine effectively inhibits the intracellular target enzymes, L-alanine racemase and D-alanyl-D-alanine synthetase involved in peptidoglycan biosynthesis (Chapter 2).

7.3.2 Fosfomycin (phosphonomycin)

This simple phosphorus-containing antibiotic (Figure 2.10) uses two different physiological transport systems to gain access to the bacterial cytoplasm:

(1) The structural resemblance of fosfomycin to α-glycerophosphate enables it to use the permease for this important biochemical. As expected, the uptake of fosfomycin is competitively inhibited by high concentrations of α-glycerophosphate.
(2) The permease for hexose 6-phosphates in certain enterobacteria and staphylococci is also exploited by fosfomycin.

Both transport systems are induced by their normal permeants, although not by fosfomycin. The therapeutic efficacy of fosfomycin in experimental infections can be enhanced by pretreatment of the infected animals with glucose 6-phosphate, which induces the bacterial permease, resulting in a higher intracellular concentration of fosfomycin. In Gram-negative bacteria fosfomycin crosses the outer membrane via nutrient channels used by α-glycerophosphate and hexose 6-phosphates.

7.3.3 Tetracyclines

Tetracycline crosses the outer membrane of Gram-negative bacteria mainly through a porin channel designated OmpF, probably as a chelation complex with a magnesium ion. Mutant cells with decreased OmpF expression exhibit some resistance to tetracycline, although not to the more lipophilic derivative, minocycline (Figure 5.7), which is thought to diffuse in its uncomplexed form across the lipid bilayer rather than through the porin channels.

It has long been known that tetracyclines are accumulated against their concentration gradients by Gram-negative and Gram-positive bacteria. This process, which is energized by the proton motive force across the cytoplasmic membrane, partially explains the antibacterial specificity of tetracyclines, since a comparable process is absent from mammalian cells, and tetracyclines, it may be

TABLE 7.2 Features of the bacterial cell envelope that influence the uptake of antibacterial agents

Envelope structure	*Effects on drug uptake*
Lipid bilayer of Gram-negative outer membrane	Retards diffusion of both water-soluble and lipophilic compounds
High-efficiency porin channels of Gram-negative bacteria	Facilitate diffusion of water-soluble molecules of molecular weight up to ~600 Da
Low-efficiency porin of *Pseudomonas aeruginosa*	Permits only slow diffusion of water-soluble antibiotics
Teichoic acids of Gram-positive bacteria	Strongly anionic character could, in principle, affect uptake of ionized molecules
Lipid bilayer of cytoplasmic membrane	Rates of passive diffusion depend on lipophilicity of permeant; permits little or no passive diffusion of water-soluble or strongly ionized molecules
Facilitated transport systems of cytoplasmic membrane	Markedy enhance rates of transfer of nutrients and structural analogues, e.g. D-cycloserine, fosfomycin; energy coupling leads to accumulation against concentration gradients
Multidrug efflux systems	Reduce intracellular drug concentrations by effluxing into external medium; occur in both wild-type and drug-resistant mutant bacteria

recalled, inhibit protein synthesis on both 70S and 80S ribosomes (Chapter 5). The energy-dependent accumulation of tetracyclines against a concentration gradient appears to have some of the hallmarks of facilitated diffusion mediated by a carrier system associated with the cytoplasmic membrane. However, no such carrier for the inward transport of tetracycline across the cytoplasmic membrane has ever been identified. Furthermore, there is scant evidence that the rate of tetracycline influx is saturable at high concentrations of the drug – a defining characteristic of carrier-mediated facilitated transport. How, then, is the energy-dependent accumulation of tetracyclines to be explained? The answer may lie in the physicochemical properties of the tetracycline molecule.

In addition to the hydrophobic character of its fused ring system, tetracycline has three ionizable centres with pK_a values of 3.3 (tricarbonyl methane group), 7.7 (phenolic β-diketone system) and 9.7 (dimethylamino group). The ratio of charged to uncharged tetracycline molecules at pH 7.4 is uncertain because of the complex ionization of the molecule. However, calculations based on the use of microscopic dissociation constants that define the protonation of tetracycline show that approximately 7% of the molecules are uncharged at physiological pH. The uncharged molecules, unlike their charged counterparts, diffuse rapidly across the cytoplasmic membrane. The proton gradient across the membrane maintains the cytoplasmic pH about 1.7 pH units higher than the pH of the external medium. The effect of the higher internal pH is to increase significantly the *fraction* of negatively charged tetracycline molecules within the cells. When equilibrium is reached by passive diffusion, the *concentration* of uncharged molecules must be the same on both sides of the membrane. The total concentration of tetracycline, i.e. uncharged plus charged molecules, is therefore higher in the bacterial cytoplasm than in the medium. Calculations based upon these assumptions predict that the intracellular concentration of tetracycline in normally metabolizing bacterial cells should be approximately four times that in the external medium. Direct measurements in *Escherichia coli*, however, reveal a 15-fold difference between the internal and external concentrations. This discrepancy is probably explained by the binding of up to 30% of the intracellular tetracycline to the ribosomes and also by a concentration of the magnesium-chelated form of tetracycline in the periplasmic space.

The maintenance of the pH gradient across the membrane depends on the energy metabolism of the cell. Compounds that collapse the gradient directly or indirectly by inhibiting energy metabolism, prevent the intracellular accumulation of tetracycline and promote the release of previously accumulated drug into the external medium.

The model described above provides the most likely explanation for the energy-dependent accumulation of tetracycline by bacterial cells. However, before the possibility of carrier-mediated influx is rejected entirely, it should be noted that tetracycline-specific, carrier-mediated efflux systems occur widely in tetracycline-resistant bacteria (Chapter 9). Furthermore, even in wild-type *Escherichia coli* there is evidence of a low-efficiency tetracycline-specific efflux system, probably involving a carrier system. The existence of these tetracycline efflux pumps therefore raises the possibility of carrier-mediated tetracycline influx.

7.3.4 Quinolones

Antibacterial quinolones (Figure 4.13), such as ciprofloxacin, have two ionizable centres, the carboxyl group, pK_a 7.5, and the 'distal' nitrogen of the piperazine ring, pK_a 6.5. Calculations similar to those applied to the tetracyclines show that at pH 7.4 approximately 10% of the molecular population is in the uncharged form. The precise values of these parameters for individual compounds depend on the nature of the substituents on the quinolone ring system.

Quinolone entry through the outer membranes of Gram-negative bacteria occurs mainly as charged molecules through the porin channels. Ionization of the carboxyl group enables quinolones to chelate with Mg^{2+} ions and a substantial part of the influx through the porin channels is probably in the form of the magnesium complex. The chelates are likely to dissociate in the more acid environment of the periplasm, leading to the establishment of a Donnan equilibrium across the outer membrane with a higher total drug concentration in the periplasm than in the external medium. This effect could explain why some quinolones are more effective against Gram-negative bacteria than against Gram-positive bacteria, because of the absence of a defined periplasmic space in the latter organisms.

Bacterial cells accumulate quinolones and it was once thought that this was another example of active drug uptake into the cytoplasm. However, much of the drug associated with Gram-negative bacteria is probably bound to surface components via magnesium chelation. Consideration of the ionization equilibria of quinolones suggests that their cytoplasmic concentration may actually be less than that in the external medium, due to the pH gradient across the cytoplasmic membrane. When equilibrium of the uncharged molecules is reached by passive diffusion across the cytoplasmic membrane, the total internal concentration (uncharged plus charged molecules) is calculated to be less than the external concentration. Interestingly, carbonyl cyanide *m*-chlorophenylhydrazone, an agent used experimentally to collapse the proton gradient across bacteria cell membranes, actually increases the uptake of certain quinolones into the cytoplasm. Abolition of the pH difference between the cytoplasm and the exterior is believed to establish a new equilibrium position by promoting a transient increased influx of quinolone that equalizes the total drug concentrations on each side of the membrane.

In summary, there is no substantial evidence for carrier-mediated accumulation of quinolones by bacteria and the observed uptake phenomena can be explained largely by the physico-chemical properties of the compounds and compartmental differences in pH within bacterial cells.

7.3.5 Sideromycins

Complex antibiotics such as ferrimycin A_1 (Figure 7.2) subvert the bacterial uptake of iron-transporting sideramines. The structural similarity of a typical sideramine, ferrioxamine B, to ferrimycin A_1 is illustrated in Figure 7.2. Sideromycins cross the Gram-negative outer membrane via the sideramine-specific porin channels and then exploit the sideramine–iron carrier in the cytoplasmic membrane to gain access to the bacterial cytoplasm. Subversion of the sideramine transport system enables sideromycins to achieve intracellular concentrations up to 100 times higher than those in the medium. The antibiotics are, however, extensively bound to intracellular sites so that the apparent internal concentration is not all due to 'free' drug. Mutations affecting the outer membrane receptors for sideramines and the cytoplasmic membrane transport mechanism both give rise to resistance to the sideromycins. Bacterial viability is not usually impaired by these mutations because most organisms have alternative systems for abstracting iron from the external environment. The ease with which bacteria develop resistance to sideromycins has precluded their clinical application.

7.3.6 Aminoglycosides

These polycationic water-soluble molecules (Figures 5.5, 5.6) have molecular weights approaching the molecular weight exclusion limit of 600 Da for porin-mediated transport through the Gram-negative outer membrane. There is uncertainty, therefore, as to the contribution this mode of transport makes to the uptake of aminoglycosides by Gram-negative bacteria. Aminoglycosides may promote their own penetration by competitive displacement of the stabilizing Mg^{2+} and Ca^{2+} ions from the LPS, thus disrupting the barrier function of the outer membrane. The initial uptake of dihydrostreptomycin by *Escherichia coli* is characterized by an electrostatic interaction between the positively charged guanidino centres and the anionic groups of LPS. Some accumulation in the periplasmic space may occur, followed by a slow, energy-dependent penetration into the cytoplasm. After 15–30 min a third phase of rapid, energy-dependent intracellular accumulation of dihydrostreptomycin begins. This final

Ferrioxamine B

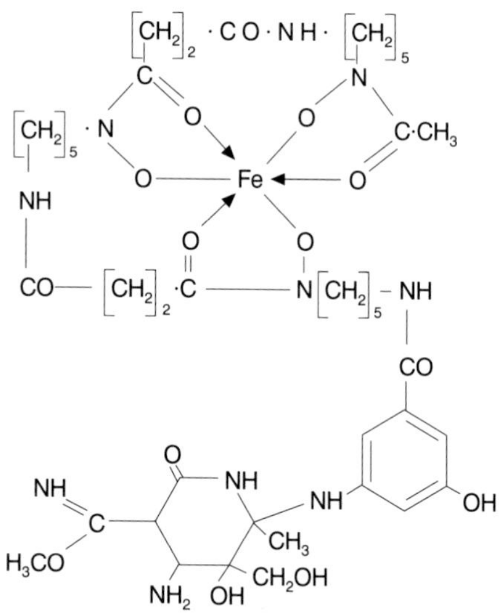

Ferrimycin A_1

FIGURE 7.2 Ferrioxamine B, an iron-chelating growth factor, and ferrimycin A_1, a structurally related antibiotic that gains access to bacterial cytoplasm via the growth factor transport system.

phase appears to be irreversible in *Escherichia coli* and the antibiotic can only be released from the cells by damaging the cell membranes with organic solvents such as toluene. The molecular mechanisms involved in the energy-dependent phases of aminoglycoside uptake are obscure. An earlier suggestion for the involvement of an aminoglycoside-inducible permease, normally responsible for accumulating polyamines, has now been discounted. The binding of aminoglycosides to 70S ribosomes further enhances their accumulation within bacteria. The intrinsic resistance of anaerobic bacteria to the aminoglycosides may be due to their limited ability to accumulate these antibiotics.

7.4 Uptake of antimicrobial drugs by eukaryotic pathogens

The mechanisms involved in the transport of drugs into fungal and protozoal pathogens have received less attention than their counterparts in bacteria. In fungi and protozoa the cytoplasmic membranes are likely to be the major permeability barrier against drug influx. The complex outer envelopes of fungi could conceivably hinder the access of some larger molecules, but in general the coarseness of the chitin and mannan meshworks probably offers little resistance to drug influx.

7.4.1 5-Fluorocytosine

This antifungal drug (Figure 4.5) provides an example of facilitated transport into fungal cells. As a close analogue of cytosine, 5-fluorocytosine is transported across the cytoplasmic membrane by the cytosine permease. Rapid intracellular metabolism of 5-fluorocytosine to several toxic pyrimidine nucleotides contributes to the maintenance of a downward concentration gradient of unchanged drug into fungal cells. A disadvantage of reliance on cytosine permease for the uptake of 5-fluorocytosine is that mutational inactivation of the transport system results in resistance to the drug.

7.4.2 Polyoxins and nikkomycins

The inhibitory activities of these peptidonucleoside antibiotics (Figure 2.20) depend on their transport into fungal cells by a peptide permease system normally intended for the accumulation of nutrient dipeptides. However, antibiotic transport on the permease system is subject to competitive inhibition by dipeptides which are commonly found in the blood and tissues of the infected host. As in the case of 5-fluorocytosine, resistance to polyoxins and nikkomycins also arises readily from mutations that inactivate the permease system.

The lipophilic character of the azole antifungal drugs (Figure 3.10) probably ensures their penetration of cytoplasmic membranes by passive diffusion. Interestingly, there is evidence in *Candida* spp. for broad-specificity efflux systems which may contribute to azole resistance by pumping drugs from the cytoplasm into the external medium.

7.4.3 Antiprotozoal drugs

With the exception of the polycationic, antitrypanosomal drug, suramin (Figure 1.2), which enters the parasite complexed with serum proteins by a process of endocytosis, most antiprotozoal drugs probably diffuse passively across the cytoplasmic membranes of their target pathogens. Positively charged antimalarial compounds, such as chloroquine (Figure 4.10), subsequently accumulate within the acidic environment of the digestive vacuoles of the parasites. Drug binding to haem released by the digestion of haemoglobin and to the haem polymer haemozoin, in the case of chloroquine and artemisinin, also contributes to the persistence of favourable chemical gradients into the vacuoles.

The antitrypanosomal drug, eflornithine (Figure 6.8), conceivably enters trypanosomes by a permease that would normally facilitate the uptake of ornithine or other polyamines. However, as yet there is no experimental evidence for this.

Further reading

Hancock, R. E. W. (1997). The bacterial outer membrane as a drug barrier. *Trends Microbiol.* **5**, 37.

Nikaido, H. (1994). Prevention of drug access to bacterial targets: permeability barriers and active efflux. *Science* **264**, 382.

Nikaido, H. (1996). Multidrug efflux pumps of Gram-negative bacteria. *J. Bact.* **178**, 5853.

Nikaido, H. and Thanassi, D. G. (1993). Penetration of lipophilic agents with multiple protonation sites into bacterial cells: tetracyclines and fluoroquinolones as examples. *Antimicrob. Agents Chemother.* **37**, 1393.

Paulsen, I. T., Brown, M. H. and Skurray, R. A. (1996). Proton-dependent multidrug efflux systems. *Microbiol. Rev.* **60**, 575.

Tute, M. S. (1972). Principles and practice of Hansch analysis: a guide to the structure–activity relationships for the medicinal chemist. *Adv. Drug. Res.* **6**, 1.

Chapter eight

The genetic basis of resistance to antimicrobial drugs

The development of safe, effective antimicrobial drugs has revolutionized medicine in the past 60 years. Morbidity and mortality from microbial disease have been drastically reduced by modern chemotherapy. Unfortunately, micro-organisms are nothing if not versatile, and the brilliance of the chemotherapeutic achievement has been dimmed by the emergence of microbial strains presenting a formidable array of defences against our most valuable drugs. This should not surprise us, since the evolutionary history of living organisms is concerned with their adaptation to the environment. The adaptation of micro-organisms to the toxic hazards of antimicrobial drugs is therefore probably inevitable. The extraordinary speed with which antibiotic resistance has spread amongst bacteria during the era of chemotherapy has been due, in large measure, to the remarkable genetic flexibility of this group of organisms.

The first account of microbial drug resistance was given by Paul Ehrlich in 1907, when he encountered the problem soon after the development of arsenical chemotherapy against trypanosomiasis. As the sulphonamides and antibiotics were brought into medical and veterinary practice, resistance against these agents began to emerge. Resistance to antibacterial and antimalarial drugs is now widespread and resistance to antifungal and antiviral drugs is of increasing concern. Our intention in this chapter is to give an outline of the genetic background to the problem of drug resistance, and in Chapter 9 we describe the major biochemical mechanisms that give rise to resistance.

The tremendous advances made in the science of bacterial genetics over the past 50 years have found a most important practical application in furthering our understanding of the problem of drug resistance. As a result we now have a fairly complete picture of the genetic factors underlying the emergence of drug-resistant bacterial populations. Although the study of the genetics of resistance in pathogenic fungi and protozoa is less advanced, recent technological developments have set the stage for significant improvements in our understanding of these organisms. The depressingly rapid emergence of drug-resistant variants of the human immunodeficiency virus (HIV) during the chemotherapy of AIDS has given a powerful impetus to the study of the genetic and biochemical basis of resistance to antiviral drugs.

The early studies on the genetics of drug resistance were bedevilled by an exhausting controversy. On the one hand were those who believed that the development of a resistant cell population could be explained by the phenotypic adaptation of the cells to an inhibitory compound without significant modification in their genotype. The opposing faction took the view that any large population of cells which was sensitive overall to a drug was likely to contain a few genotypically resistant cells. The continued presence of the drug resulted in the expansion of the numbers of resistant cells by a process of selection.

Evidence gathered over the years strongly supports the second of these two theories. As we shall see, there are examples of phenotypic adaptation to antimicrobial drugs but such cells are usually genotypically different from wild-type cells. When the selective pressure applied by an antimicrobial drug is removed, the resistant microbial population may revert to drug sensitivity if the resistant cells are at a selective disadvantage to drug-sensitive cells in a drug-free environment and could therefore eventually be outgrown by the sensitive cells.

8.1 Mutations and the origins of drug-resistance genes

Once it was accepted that drug-resistant organisms were genetically different from the wild types it was natural to consider how such differences might arise. One obvious possibility is that of spontaneous mutations. These can arise is several ways:

1. damage to the genome caused by adverse environmental factors, including ionizing radiation and chemical mutagens;
2. base-pairing errors during genomic replication; and
3. promiscuous intragene insertion of extraneous genetic material, such as transposons, that corrupt the correct flow of information from the wild-type genome.

Spontaneous mutations are relatively rare, of the order of one mutation per 10^7–10^{11} cells per generation, although in organisms lacking a proof-reading mechanism during genomic replication, as in HIV, the mutation rate can be much higher. When the vast numbers of organisms in microbial populations are considered, the probability of even low mutation rates causing drug resistance is quite high. The simple and elegant technique of replica plating convincingly demonstrates that spontaneous mutations to drug resistance can occur in drug-sensitive bacterial populations in the absence of drugs (Figure 8.1). A spontaneous mutation may occasionally cause a large increase in resistance, but resistance often develops as a result of numerous mutations, each giving rise to a small increase in resistance. In this situation highly resistant organisms emerge only after prolonged or repeated exposure of the microbial population to the drug.

8.1.1 Spontaneous mutations and drug resistance in HIV

A major challenge to the effective treatment of AIDS is the alarming speed with which HIV becomes resistant to the various drugs deployed against it, including inhibitors of viral reverse transcriptase and HIV protease. The origins of drug resistance in HIV lie in the high rate of viral replication and the ease with which spontaneous mutations arise in its RNA genome. As with other single-stranded RNA viruses, HIV lacks a proof-reading mechanism to eliminate any sequence errors that may occur during nucleic acid replication and mutations occur with high frequency. During the course of an infection the combination of high replication and mutation rates permits rapid and extensive evolution of the viral population in response to immunological and chemotherapeutic challenges to its survival. For example, within weeks of starting treatment with the reverse transcriptase inhibitor, lamivudine, spontaneous mutation results in the replacement in the reverse transcriptase of the circulating viruses of methionine-184 by valine, a change associated with high-level resistance to lamivudine.

In contrast with lamivudine resistance, the emergence of resistance to other reverse transcriptase inhibitors, such as AZT and nevirapine, and to inhibitors of HIV protease reveals a more varied pattern of mutations. These differences are probably associated with the levels of resistance conferred by individual mutations. The replacement of methionine-184 by valine in lamivudine-resistant HIV confers a marked selective advantage over viruses with less effective mutations and results in an apparently uniform population of drug-resist-

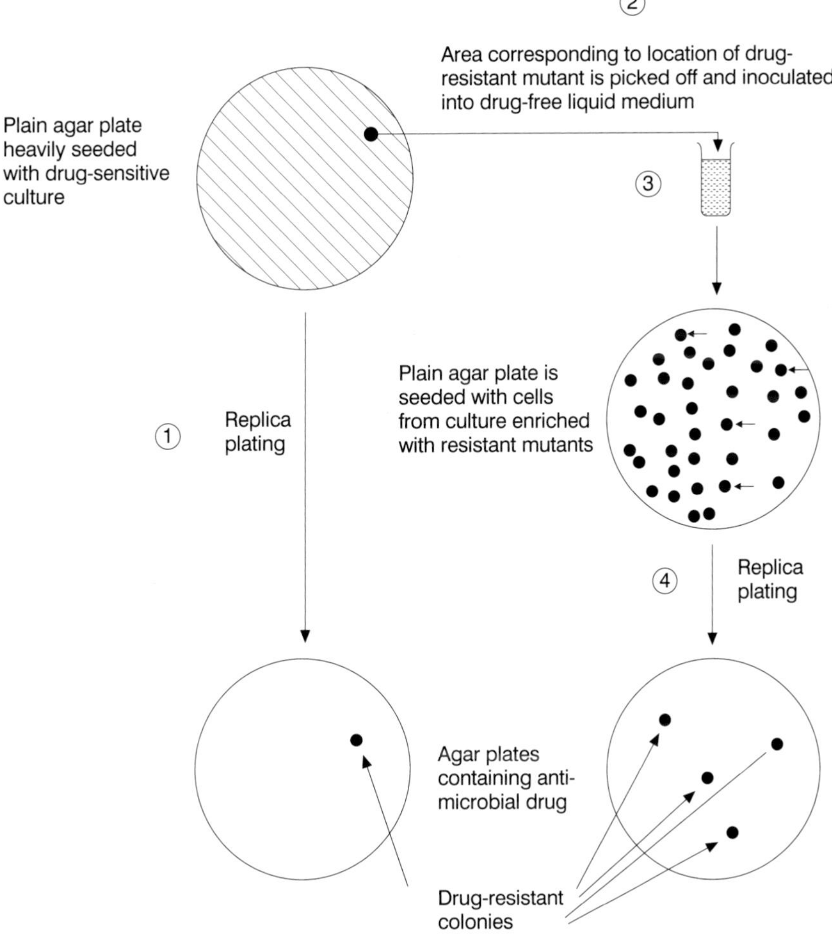

FIGURE 8.1 The technique of replica plating reveals the existence of drug-resistant cells in a population that is overall drug-sensitive. A plain agar plate is heavily seeded with cells from the drug-sensitive culture and is incubated until growth occurs. Cells are transferred by a velvet pad to a plate containing the antibacterial drug: this plate is then incubated and the position of any colonies noted. The area on the drug-free plate corresponding to the location of the resistant colony on the drug plate is picked off and cultured in drug-free medium. Although still contaminated with sensitive cells, this culture will contain many more resistant cells than the original culture. Plating out of the 'enriched' culture on a plain plate followed by replication to a drug plate therefore reveals a higher incidence of drug-resistant colonies. The experiment shows that drug-resistant mutants occur in a bacterial population not previously exposed to the drug.

ant viruses. In contrast, the balance of advantage of the various mutations causing resistance to the other drugs may be less clear-cut, so that, depending upon differences in the nature of HIV infections in individual patients, distinct populations of viruses resistant to the same drug may emerge with a variety of different mutations. The clinical approach to coping with the rapid acquisition of drug resistance by HIV is to treat patients with a combination of drugs. In this way the emergence of resistant viruses can be delayed by months or even years.

At one time it was believed that spontaneous mutations followed by the selection of resistant organisms in the presence of drug, provided a general explanation for the emergence of all drug-resistant populations. However, while this appears to be true in the case of HIV, the realization that bacteria can acquire additional genetic material by conjugation, transformation and transduction led to the conclusion that spontaneous mutations make only a relatively minor contribution to the emergence and spread of drug resistance in bacteria. However, spontaneous mutations do cause resistance to wholly synthetic antibacterials, such as the quinolones and the resistance of *Mycobacterium tuberculosis* to commonly used antituberculous drugs. Spontaneous mutations also lead to the progressive modification of drug-inactivating enzymes, thus enabling them to cope with novel chemical variants of the original drug molecule. This is well established in the case of the β-lactamases (Chapter 9).

8.1.2 Origin of clinically important resistance genes in bacteria

As we shall see in the next chapter, the biochemical machinery conferring bacterial resistance to drugs of major importance in medicine may be quite complex. Understandably, therefore, there is considerable interest in the origins of the genes that encode drug-inactivating enzymes, drug efflux pumps and enzymes that depress drug sensitivity by the covalent modification of drug targets. Antibiotic-producing bacteria, such as streptomycetes, for example, protect themselves against the toxic effects of their own antibiotics, with enzymes that specifically inactivate compounds such as aminoglycosides, chloramphenicol and β-lactams. In addition, many streptomycetes express β-lactamases even though they do not produce β-lactams, presumably as a protective measure against β-lactams synthesized by other organisms in the micro-environment. Pumped efflux of tetracycline and the enzymic modification of ribosomal RNA associated with resistance to erythromycin have both been identified in streptomycetes. A comparison of nucleic acid and protein sequence data supports the possibility that the genes for aminoglycoside-inactivating enzymes found in aminoglycoside-resistant clinical isolates may have originated from streptomycetes. In general, however, evidence for a streptomycete origin of drug-resistance genes in pathogenic bacteria is limited.

Mosaic genes

Although bacterial genes that encode antibiotic inactivating enzymes and drug efflux pumps almost certainly evolved hundreds of millions of years ago, there is evidence for the evolution of other modes of antibacterial drug resistance during the modern era of chemotherapy. A remarkable example of this is the emergence of mosaic genes through the process of interspecies genetic recombination. By far the most common mechanism of resistance to β-lactam antibiotics is that of antibiotic hydrolysis by β-lactamases, which are probably of ancient origin. However, β-lactam resistance in several important pathogens, including *Haemophilus influenzae, Neisseria gonorrhoeae, Streptococcus pneumoniae, Staphylococcus aureus* and *Staphylococcus epidermidis*, can be due to penicillin-binding proteins (PBPs; Chapter 2) with reduced affinity for β-lactams. This type of resistance is relatively rare because the killing action of β-lactams depends on drug interactions with several high molecular weight PBPs and resistance therefore necessitates reductions in β-lactam affinity in each PBP. Although it is conceivable that such reductions in affinity could have arisen gradually from incremental changes in protein structure due to the accumulation of mutations in the PBP genes, it is now clear that recombination amongst PBP genes from different species is a major cause of the low-affinity PBP phenotype in bacteria that are readily transformable by DNA released from lysed bacteria.

Analysis of the sequences of genes for PBP2 from penicillin-sensitive and penicillin-resistant meningococci and gonococci reveals that whereas the sequences from penicillin-sensitive bacteria are uniform, the resistant gene sequences have a mosaic structure. The mosaics are made by the recombination of regions essentially identical with those from penicillin-sensitive bacteria with regions with significantly divergent sequences. The mosaic genes encode PBP2 variants with decreased affinity for penicillin. Sequence information obtained from bacterial DNA databases shows that the divergent regions in the mosaic genes originate from *Neisseria flavescens* and *Neisseria cinerea*. PBP2 prepared from specimens of *Neisseria flavescens* preserved from before the antibiotic era has much lower affinity for penicillin than PBP2 from either *Neisseria gonorrhoeae* or *Neisseria meningitidis*. The mosaic genes are thought to have arisen by interspecies recombination, made possible by the ease of genetic transformation by DNA released from lysed cells (see below) amongst these bacteria.

Mosaic genes encoding low-affinity PBPs 1a, 2x, 2b and 2a have been isolated from *Streptococcus pneumoniae* resistant to both penicillins and cephalosporins. Pneumococci are also readily transformable and the divergent regions of the mosaic genes appear to have originated from several other bacterial species.

An important example of a low-affinity PBP that does not appear to be the result of mosaic gene formation is PBP2a encoded by the *mecA* gene that is responsible for the notorious methicillin-resistant *Staphylococcus aureus* (MRSA). This organism is not naturally transformable by DNA and the *mecA* gene is carried by a transposon which integrates into the bacterial chromosome. The origin of the *mecA* gene remains unknown at present.

Resistance to sulphonamides amongst meningococci also resulted from interspecies recombination yielding mosaic genes that encode sulphonamide-resistant dihydropteroate synthase (Chapter 4). Allelic variations in the *tetM* gene, which determines the ribosome protection form of resistance to tetracycline (Chapter 9), are due to recombination amongst the distinct *tetM* alleles found in *Staphylococcus aureus* and *Streptococcus pneumoniae,* although it is not clear to what extent mosaic *tetM* genes contribute to clinically significant bacterial resistance to tetracyclines. Finally, it should be noted that the generation of mosaic genes by interspecies recombination in bacteria is not limited to resistance genes. The phenomenon is widespread in bacteria and underlies, for example, the highly divergent genes that encode the proteins of the outer membranes of *Neisseria* spp.

8.2 Gene mobility and transfer in bacterial drug resistance

The alarming spread of drug resistance amongst bacterial pathogens owes much to the remarkable facility of bacteria to mobilize genes in both chromosomal and plasmid DNA and to transfer and exchange genetic information. The realization that drug resistance could be transferred from resistant to sensitive bacteria came from combined epidemiological and bacterial genetic studies many years ago in Japan. The first clue was provided by the isolation, from patients suffering from dysentery, of strains of shigella resistant to several drugs, including sulphonamides, streptomycin, chloramphenicol and tetracycline. Even more striking was the discovery that both sensitive and multiresistant strains of shigella could occasionally be isolated from the same patient during the same epidemic. Most patients harbouring multiresistant shigella also had multiresistant *Escherichia coli* in their intestinal tracts. This suggested that drug resistance markers might be transferred from *Escherichia coli* to shigella and vice versa. Subsequently it was confirmed that Gram-negative bacteria can indeed transfer drug resistance not only to cells of the same species but also to bacteria of different species and genera. However, before we describe the transfer of drug-resistance genes between bacterial cells we must first consider the movement of genes within the bacterial genome itself.

8.2.1 Transposons

For many years the movement of genes among plasmids and chromosomes was believed to result from classical recombination, dependent on the product of the bacterial *recA* gene and the reciprocal exchange of DNA in regions of considerable genetic homology. This permits the exchange of genetic information only between closely related genomes. However, such a restricted phenomenon seemed unlikely to explain the widespread distribution of specific resistance determinants. It is now clear that the acquisition of genetic material by plasmids and chromosomes in both Gram-negative and Gram-positive bacteria is not limited by classical *recA*-dependent recombination. Replicons, known as transposons, are able to insert themselves into a variety of genomic sites having no common ancestry, that is, homology with the inserting sequence. In the simplest transposons the whole of the genetic information is concerned with the insertion function. Insertion sequences (IS elements) are sequences of approximately 750–1600 base pairs bounded at each end by inverted repeats of 15–20 base pairs that are characteristic of individual transposons. The IS elements harbour a gene for a specific endonuclease, called transposase. The genes for drug resistance are carried by more complex, or composite, transposons designated by the prefix Tn. The central coding region contains the genes for the transposase and also for resolvase, an enzyme that catalyses recombination between two insertion sequences. These genes, together with those for drug resistance, are again bounded by short inverted repeats (Figure 8.2). Many types of composite transposon have been identified in both Gram-positive and Gram-negative bacteria, carrying arrays of drug-resistance genes.

In some transposons the drug-resistance genes are arranged within structures called integrons. These consist of an *int* gene that encodes for a site-specific recombination enzyme or integrase, an integron receptor site, *attI*, and one or more gene cassettes. Usually each gene cassette contains a single drug-resistance gene and a specific recombination site, called a 59-base-pair element, located downstream of the gene. The association of the integrase function with the specific recombination site confers mobility on gene cassettes and the ability of integrons to capture and integrate whole arrays of cassettes. There are multiresistance integrons that confer various combinations of resistance to β-lactams, aminoglycosides, trimethoprim, chloramphenicol, antiseptics and disinfectants. More than 40 gene cassettes and three classes of integrons are known and the reader is referred to reviews listed under 'Further reading' for detailed descriptions of this complex field. Although integrons are found in transposons, they also occur frequently as independent entities.

Transposons and integrons confer two distinct modes of gene mobility. Mobilization of an entire transposon along with its complement of drug-resistance genes occurs by a process of replicative transposition. Typically, a replicated copy of the transposon can insert virtually anywhere within the bacterial genome. The transposase introduces staggered cuts, nine base pairs apart, at the donor site in the transposon and at the intended recipient site. The recipient site, of 4–12 base pairs in length, is then replicated to form non-inverted repeats on either side of the inserted transposon. In contrast, the process of gene cassette excision and capture in integrons is accomplished by a quite different process called site-specific recombination. The integrase associated with one or more cassettes catalyses insertion by promoting recombination between the 59-base-pair element and the *attI* site of the integron. Cassette excision is the reverse of integration and generates a circularized form of the cassette which may exist independently for extended periods. Both cassette excision and insertion normally occur only at the specific *attI* site, although there are rare examples of cassettes that integrate into non-specific sites.

To summarize, therefore, drug-resistance genes in bacteria are subject to two major modes of intragenic mobilization that promote a continual flux of resistance determinants around bacterial DNA:

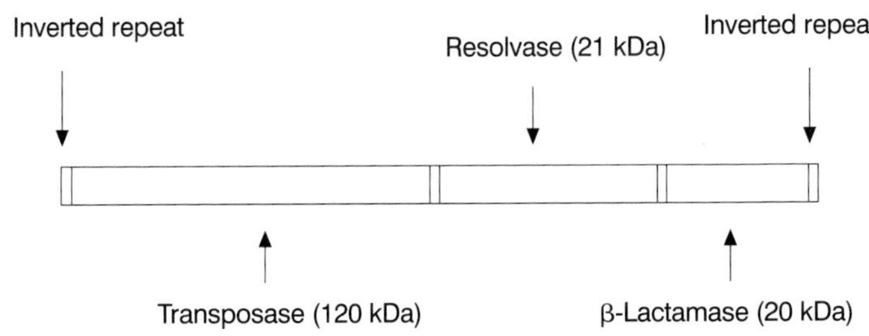

FIGURE 8.2 The structure of a composite transposon (Tn3).

1. Resistance genes associated with transposons, whether or not as cassettes, are mobilized along with the rest of the transposon and can be inserted essentially anywhere in the bacterial genome, either chromosomal or plasmid, by means of replicative transposition.
2. Both transposon-associated and independent integrons containing resistance gene cassettes exchange and capture cassettes by site-specific recombination.

Conjugative transposons

The transposons described so far are, by themselves, unable to promote gene transfer by conjugation between bacterial cells, although they participate as passengers during R-plasmid transfers. However, other mobile DNA elements, referred to as conjugative transposons, encode not only drug-resistance genes but also proteins that activate intercellular conjugation and gene transfer. Conjugative transposons occur very widely in Gram-positive bacteria and contribute to the spread of drug resistance among major pathogens such as *Streptococcus* spp. and *Enterococcus* spp. Currently *Bacteroides* spp. are the only Gram-negative bacteria known to harbour conjugative transposons under natural conditions, although in the laboratory they can be transferred from Gram-positive organisms to *Escherichia coli* and *Neisseria* spp.

The first conjugative transposon to be discovered was Tn916, which carries resistance to tetracycline. Originally Tn916 was detected on the chromosome of a multiresistant isolate of *Enterococcus* (previously *Streptococcus*) *faecalis* but it was observed to integrate readily into plasmids and into many sites of the chromosomes of bacterial recipients of Tn916. A closely related conjugative transposon, Tn1545 found in *Streptococcus pneumoniae,* also mediates tetracycline resistance as well as resistance to erythromycin and kanamycin. Currently there are 10 members of the Tn916–Tn1545 family and conjugative transposons have been identified in 52 species of bacteria.

Strictly speaking, conjugative transposons are not 'true' transposons since their mechanism of transposition differs significantly from that of the latter group. Although the base sequence of Tn916 is known, the functional assignment of the 24 open reading frames (ORFs) is at an early stage. Two of the ORFs, *orf1* and *orf2*, have sequences related to those of the excisase and integrase functions of lambdoid bacteriophages and are therefore assumed to be involved in the process of transposon excision and integration. The sequence of events during the mobilization of conjugative transposons is as follows:

1. Staggered cuts are introduced at each end of the transposon, leaving 6-nucleotide, single-stranded stretches of DNA, known as coupling sequences.
2. The non-complementary coupling sequences are then ligated to generate covalently closed, double-stranded circular intermediates.
3. During the insertion stage the coupling sequences form temporary non-base-pairing interactions with the target site. It is not yet clear how correct base pairing is subsequently established but it may involve either replication through the insertion region or mismatch

repair. The insertion process of conjugative transposons differs from that of 'true' transposons in that the target site is not duplicated.

The intercellular conjugation process promoted by conjugative transposons is not well understood. Unlike the process mediated by R-plasmids in Gram-negative bacteria (see below) surface pili do not appear to be involved. The analysis of the genes deployed during conjugative transposon transfer from one cell to another continues and the reader is referred to reviews on this topic listed at the end of the chapter. It is clear that only single-stranded copies of the transposons are transferred to recipient cells during conjugation.

The intercellular traffic of conjugative transposons is highly regulated but only in the case of bacteroides transposons is the regulatory signal known. Bacteroides transposons all carry the genes for a form of tetracycline resistance, *tetM,* dependent upon ribosomal protection and it is remarkable to find that tetracycline is a highly effective stimulant of conjugative transposon mediated mating in these species. Certain transposon genes activated by tetracycline promote conjugation and the transfer of bacteroides transposons, and also of R-plasmids sharing the same donor cells. *Bacteroides* spp. containing conjugative transposons therefore have an extraordinarily effective defence system against tetracycline because the antibiotic actually stimulates the spread of the resistance genes throughout the bacterial population.

8.2.2 R-plasmids

Cellular conjugation, mediated by R-plasmids, is the major mechanism for the spread of drug resistance through Gram-negative bacterial populations. R-plasmids are usually separate from the bacterial chromosome. They consist of two distinct but frequently linked entities:

1. the genes that initiate and control the conjugation process; and
2. a series of one or more linked genes, often found within transposon–integron complexes, that confer resistance to specific antibacterial agents.

The conjugative region is closely related to the F-plasmid which also confers on Gram-negative bacteria the ability to conjugate with cells lacking an F-plasmid. A complete R-plasmid resembles the F-prime plasmid (F′) in carrying genetic material additional to that which controls conjugation.

An extraordinary variety of R-plasmids has been described, carrying various combinations of drug-resistance determinants. Apart from the property of drug resistance, the other phenotypic characteristics conferred on the cells by different R-plasmids have prompted attempts at classification. The characteristics include:

1. the ability (fi^+) or inability (fi^-) to repress the fertility properties of a co-resident F-plasmid in the same cell;
2. the type of sex pilus (see below) that the R-plasmid determines;
3. the inability of R-plasmids to coexist in a bacterium with certain other plasmids, which permits the division of R-plasmids into several incompatibility groups;
4. the presence of genes in the R-plasmid that specify DNA restriction and modification enzymes.

An R-plasmid is not defined on the basis of a single characteristic but rather on a combination of properties.

Molecular properties of R-plasmids

R-plasmids can be isolated from host bacteria as circular DNA (Figure 8.3) in both closed and nicked forms, and both forms coexist in the cell. The closed circular structure is probably adopted by R-plasmids not engaged in replication. The contour lengths and thus molecular weights of isolated R-plasmids depend very much on the host bacterium and upon the culture conditions prevailing immediately before the isolation procedure. The R-plasmid may sometimes dissociate into its conjugative and resistance determinants.

This is more common in some host species, e.g. *Proteus mirabilis* and *Salmonella typhimurium*, than in *Escherichia coli* where dissociation is rare. Dissociation seems to depend on the activity of a simple transposon that may be inserted at the junction of the two regions. The molecular weights of between 50×10^6 and 60×10^6 kDa of the conjugative regions from R-plasmids are much greater than those of the drug-resistance genes. For example, the resistance determinants for chloramphenicol, streptomycin, spectinomycin and sulphonamide have a combined molecular weight of only 12×10^6 kDa.

The number of R-plasmids harboured by individual bacteria is determined by the properties of the plasmid and their hosts as well as by the culture conditions. As a general rule (to which there are exceptions), the larger R-plasmids are present only in a limited number of copies (between one and four) per chromosome in *Escherichia coli*, whereas in *Proteus mirabilis* the number is much more variable and even varies during the growth cycle. Conditions that give rise to an increased number of R-plasmid copies are sometimes associated with enhanced resistance. However, the level of cellular resistance does not always reflect the number of resistance gene copies. For example, although the number of R-plasmid copies is frequently greater in *Proteus mirabilis* than in *Escherichia coli*, the level of resistance to several drugs expressed in the former organism is usually lower than in *Escherichia coli*.

Cellular conjugation and R-plasmid transfer

Cells bearing an R-plasmid (R^+) are characterized by their ability to produce surface appendages known as sex pili. The sex pili of R^+ bacteria closely resemble those produced by F^+ organisms.

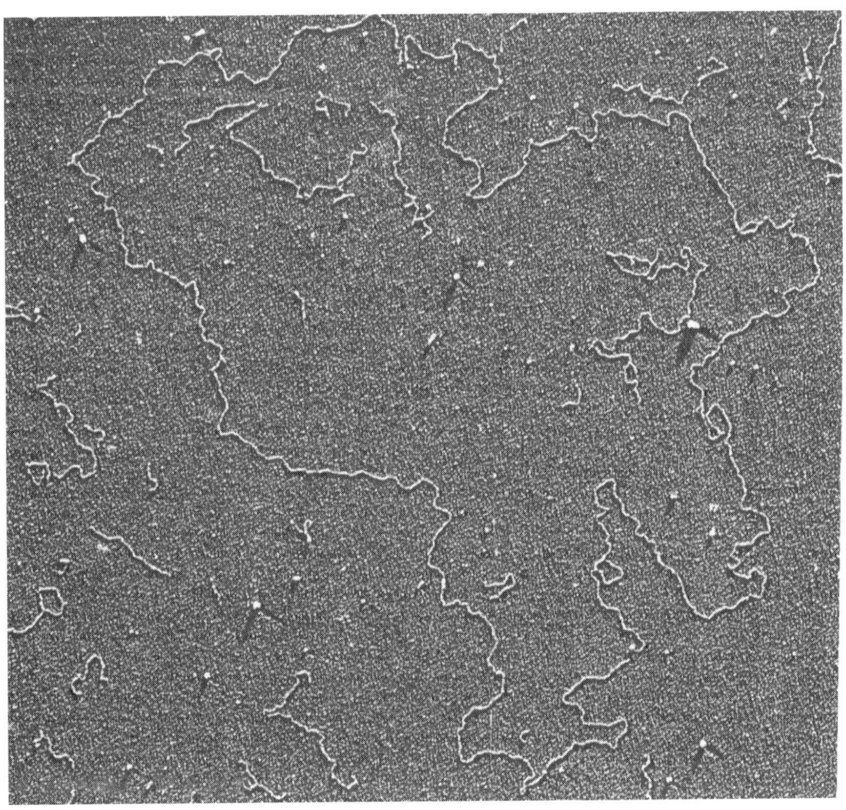

FIGURE 8.3 Electron micrograph of R-plasmid DNA isolated from *Proteus mirabilis* harbouring an R-plasmid with resistance markers to streptomycin, sulphonamides and chloramphenicol. The circular DNA has a total length of 28.5 μm. (This photograph is reproduced by kind permission of Dr Royston Clowes and the American Society for Microbiology: *J. Bacteriol.* 97, 383 (1969).)

When R^+ cells are mixed with sensitive R^- cells mating pairs are immediately formed by surface interaction involving the sex pili. The transfer of a copy of the R-plasmid from the R^+ to the R^- cell begins and the acquisition of the R-plasmid by the recipient cell converts it to a fertile, drug-resistant cell that can in turn conjugate with other R^- cells. In this way drug resistance spreads rapidly through the bacterial population. Intensive study of the process of R-factor- and F-factor-mediated cellular conjugation in Gram-negative bacteria has revealed many of the details of the process. What may appear superficially to be a fairly simple phenomenon is, in fact, highly complex, and here we provide only an outline of the process. A fuller account of the current understanding of Gram-negative conjugation can be found in a review listed under 'Further reading'.

The conjugal pair is brought into close surface contact by the attachment of the pilus of the donor cell to the recipient and its subsequent retraction, i.e. a process of 'reeling in'. The interaction between the cells triggers cleavage of a specific strand of the donor R-plasmid in the origin-of-transfer site (*oriT*) within a protein–DNA complex called the relaxosome, which contains the strand cleaving, or relaxase, enzyme. Only one strand, the T-strand, which is unwound following plasmid cleavage at the *oriT*, is transferred in a 5′ to 3′ direction from the donor to the recipient cell. The routes of the DNA strand out of the donor cell and into the recipient cell are both unknown. Pores formed by a fusion process between the inner, cytoplasmic and outer membranes may be involved, and a contribution of the sex pilus has long been surmised but remains unproven. Once in the recipient cell, the ends of the transferred strand are ligated to produce covalently closed circular DNA. Finally DNA replication, catalysed by DNA polymerase III, generates double-stranded plasmid DNA from the single-stranded molecules in both donor and recipient cells.

Fortunately perhaps, the frequency of R-plasmid transfer is much lower than that of F transfer. After infection of an R^- cell with an R-plasmid a repressor accumulates which eventually inhibits sex pilus formation. The ability to conjugate is therefore restricted to a short period immediately after acquisition of the R-plasmid. Sex pilus production in F^+ cells, by contrast, is not under repressor control and conjugal activity is therefore unrestricted. Mutant R-plasmids without the ability to restrict sex pilus formation exhibit a much higher frequency of R-plasmid transfer.

It is also worth noting that certain R-plasmids, and other self-mobilizing plasmids without drug-resistance genes, can promote the intercellular transfer of co-resident plasmids that lack the genetic information for conjugation and transfer. The latter mobilizable plasmids achieve transfer either by using the conjugal apparatus furnished by the self-mobilizing plasmids (*trans* mobilization) or by integration with these plasmids (*cis* mobilization). Clearly, such co-operative interactions amongst plasmids add significantly to the genetic flexibility of bacteria and to their ability to spread drug resistance through microbial populations.

Clinical importance of R-plasmids

It is generally agreed that R-plasmids existed before the development of modern antibacterial drugs. Clearly though, the widespread use and abuse of these drugs led to a vast increase in the incidence of drug resistance caused by R-plasmids. This has been especially noticeable in farm animals, which in many countries receive clinically valuable antibacterial drugs, or compounds chemically closely related to them, in their foodstuffs as growth enhancers. The animals act as a reservoir for Gram-negative bacteria, such as *Escherichia coli* and *Salmonella typhimurium,* harbouring R-plasmids potentially transferable to man. Fortunately, some countries have restricted the growth-enhancer application of clinically valuable antibiotics, although contravention of the regulations is not unknown.

The adverse contribution of R-plasmid-mediated drug resistance to human morbidity and mortality is undeniable. For example, the major requirement in the treatment of neonatal diarrhoea caused by certain pathogenic strains of *Escherichia coli* (a potentially dangerous condition) is the prevention of fatal dehydration. Even so, elimination of the pathogenic organisms may also be important, but this is often difficult in the face of multiple resistance to commonly used antibacterial agents. In one notorious outbreak the children were infected with a pathogenic strain of *Escherichia coli* resistant to β-lactams, streptomycin, neomycin, chloramphenicol and tetracyclines. The infection eventually responded to gentamicin, which was the only drug of those tested to which the pathogenic bacteria were sensitive. Another potentially alarming development has been the appearance of the typhoid organism, *Salmonella typhi* carrying an R-plasmid with genes for resistance to chloramphenicol and cotrimoxazole, the drugs most commonly used to treat this disease.

Certain ecological factors probably limit the clinical threat posed by R-plasmids. In the environment of the gastrointestinal tract the conjugal activity of R^+ bacteria may be less than that in the ideal culture conditions of the laboratory. The emergence of an R^+ population of bacteria during antibiotic therapy is more likely to result from selection of resistant cells than from extensive conjugal transfer of resistance. After cessation of antibiotic treatment, the numbers of R^+ bacteria in the faeces fall, although usually not to zero. Low-level antibiotic contamination of the environment and/or a previously unsuspected persistence of drug-resistance and resistance-transfer genes in bacterial populations may contribute to this potentially serious situation.

8.2.3 Conjugative transfer of drug-resistance genes between Gram-positive and Gram-negative bacteria

Although conjugative transfer of resistance genes from Gram-negative to Gram-positive bacteria can be demonstrated in the laboratory using specifically constructed R-plasmids, it is not clear to what extent the transfer of R-plasmids to Gram-positive bacteria occurs naturally. However, the conjugative transposon Tn916 mediates conjugative resistance gene transfer from *Escherichia coli* to several Gram-positive species, including *Enterococcus faecalis*.

Plasmids from Gram-positive bacteria are generally not stably maintained in Gram-negative cells and the possibility of conjugal transfer of resistance genes from Gram-positive to Gram-negative bacteria in clinical situations has been difficult to establish. However, the laboratory observation of the conjugative transfer, albeit at low frequency, of a specifically constructed hybrid plasmid from *Enterococcus faecalis* to *Escherichia coli* suggests that such transfers might occur naturally in environments such as the intestinal tracts of patients treated with antibiotics. A clinical finding in support of this possibility is the detection of a gene for erythromycin resistance, originating from Gram-positive cocci, in Gram-negative enterobacteria from patients undergoing erythromycin therapy.

8.2.4 Non-conjugal transfer of resistance genes

Transduction

During the two distinct processes of phage transduction, which occur in both Gram-positive and Gram-negative bacteria, genetic information is transferred by phage particles from one bacterium to a related phage-susceptible cell.

Generalized transduction may occur during the lytic phases of both virulent and temperate phages. Fragments of degraded host DNA, both chromosomal and plasmid, harbouring drug-resistance determinants, recombine with phage DNA and become packaged into newly generated phage particles. Lytic release of the phages enables them to inject their recombinant DNA into other bacteria. However, this does not initiate another round of lysis but rather the integration of some of the phage DNA carrying a drug-resistance gene into the recipient bacterial genome. The integrated genes are thenceforth replicated and expressed in the recipient bacteria. In abortive transduction the non-integrated, transduced DNA survives and replicates as a plasmid. The drug-resistant phenotype is maintained in both types of transduction.

The process of specialized transduction is quite distinct and depends on an error in the lysogenic cycle. Excision of the phage DNA from the host genome during induction of the lytic phase is insufficiently precise and carries some of the bacterial DNA along with phage DNA. The resulting phage genome contains up to 10% of the bacterial DNA next to the phage integration site in the bacterial genome. Clearly this process has the potential for generating infectious phage particles that carry bacterial genes for drug resistance. Although recombinant or defective phages arising from specialized transduction are able to inject their DNA into new hosts, they cannot reproduce independently, nor are they lysogenic. Specialized transduction has been most thoroughly studied with lambda phage and the reader is referred to a relevant text on bacterial genetics (listed in 'Further Reading') for a detailed account of the mechanisms involved in lambda phage transduction. In general terms, relatively little of the injected recombinant phage DNA is integrated into the bacterial genome unless the phage population contains normal phages as well as the defective phages. The normal phages insert into the bacterial genome at a specific *att* site that resembles the phage *att* site. The insertion process generates two hybrid bacterial–phage *att* sites where the defective recombinant phage DNA can insert. The presence of the normal phage renders the bacteria very susceptible to the induction of phage-mediated lysis. The resulting lysate, containing roughly equal amounts of both defective, recombinant phages and normal phages, is highly efficient in transduction.

While transduction of drug-resistance determinants is readily demonstrated under laboratory conditions, its contribution to the spread of drug resistance in natural and clinical settings is difficult to quantify.

Transformation

Under certain conditions most genera of bacteria can absorb, integrate and express fragments of 'naked' DNA containing intact genetic information, including that for drug resistance. The phenomenon of transformation of bacteria by DNA is more complex than it may appear at first glance. It has been most thoroughly investigated in the species in which it was first discovered more than 50 years ago, *Streptococcus pneumoniae.* Only bacteria in a state of competence are able to absorb and integrate exogenous DNA into their own genome. *Streptococcus pneumoniae* becomes competent during exponential growth when the population density exceeds 10^7–10^8 cells ml^{-1}. Under these conditions the bacteria secrete a competence factor that stimulates the synthesis of up to 10 other proteins essential for transformation. Competent cells bind double-stranded DNA provided that its molecular weight is at least 500 kDa. One strand of the DNA is hydrolysed by an exonuclease associated with the cell envelope, and the remaining strand enters the cell while bound to competence-specific proteins. Integration into a homologous region of the recipient genome probably occurs by the process of non-reciprocal general recombination.

Whereas competent *Streptococcus pneumoniae* can take up DNA from a range of bacterial species, the important Gram-negative opportunist pathogen *Haemophilus influenzae* is more fastidious and only accepts DNA from closely related species. Furthermore, *Haemophilus influenzae* does not produce a competence factor but absorbs double-stranded DNA encapsulated in membrane vesicles. Although transformation is a widespread phenomenon, it is not surprising to find important differences in the details of the actual mechanism among the bacterial species. The complexity and diversity of the transformation system indicates its evolutionary importance in the exchange of genetic information in the bacterial world. The frequency of transformation of genetic markers under laboratory conditions can be as high as 10^{-3}, that is, one cell in every thousand takes up and integrates a particular gene. Transformation is probably therefore a significant contributor to the spread of drug-resistance genes during the antibiotic era. A specific example of the relevance of transformation to drug resistance is illustrated by the existence of the mosaic genes for PBPs with diminished affinity for β-lactam antibiotics, for sulphonamide-resistant dihydropteroate synthase and for the *tetM* form of tetracycline resistance referred to previously. Unfortunately, the extent to which transformation occurs in relevant environments, such as hospital wards and the intestinal tract, cannot be quantified with any certainty.

8.3 Global regulators of drug resistance in Gram-negative bacteria

We have seen how the capture of several determinants for resistance to individual drugs by mobile and transferable genetic elements can result in bacteria acquiring the multidrug-resistant phenotype. However, Gram-negative bacteria have yet another means of achieving a similar end. Genes called regulons exert transcriptional control over several distinct chromosomal genes that influence resistance to a diverse range of antibiotics. The *marA* locus in *Escherichia coli* comprises an operon, *marRAB*, whose expression is inducible by at least two antibiotics, tetracycline and chloramphenicol. Resistance to these and to many other drugs is increased by the induction process. The MarA protein encoded in the operon belongs to a family of transcriptional regulators and controls the expression of numerous other genes, probably in concert with MarR and MarB proteins. Mar-regulated genes include those encoding drug efflux systems and the outer-membrane porin protein, OmpF. Enhanced drug efflux combined with diminished outer membrane permeability caused by a reduction in the expression of OmpF underlie the simultaneous increase in resistance to a range of structurally unrelated drugs that is mediated by the *marA* locus.

Another global regulator of drug resistance, *ramA,* has been found in the important Gram-negative pathogen *Klebsiella pneumoniae.* The *ramA* gene encodes a transcriptional activator protein, RamA, that is distantly related to the MarA protein of *Escherichia coli.* Like MarA, RamA confers resistance to a wide range of structurally unrelated drugs by upregulating the expression of several genes. RamA-mediated resistance also appears to depend upon a combination of drug efflux and a reduction in the level of the OmpF protein in the outer membrane.

Genetic loci resembling *marA* and *ramA* are widespread among Gram-negative bacteria, although the levels of resistance they confer are relatively modest compared with those mediated by the 'classical' resistance genes. Nevertheless, the phenomenon of multidrug resistance determined by global regulatory genes could be a significant contributor to the overall problem of drug resistance in Gram-negative bacteria.

Further reading

Alekshun, M. N. and Levy, S. B. (1997). Regulation of chromosomally mediated multiple antibiotic resistance: the *mar* regulon. *Antimicrob. Agents Chemother.* **41**, 2067.

Courvalin, P. (1994). Transfer of antibiotic resistant genes between Gram-positive and Gram-negative bacteria. *Antimicrob. Agents Chemother.* **38**, 1447.

George, A. M., Hall, R. M. and Stokes, H. W. (1995). Multidrug resistance in *Klebsiella pneumoniae*: a novel gene *ramA* confers a multidrug resistance phenotype in *Escherichia coli. Microbiol.* **141**, 1909.

Lanka, E. and Wilkins, B. M. (1995). DNA processing reactions in bacterial conjugation. *Ann. Rev. Biochem.* **64**, 141.

Leigh Brown, A. J. and Richman, D. D. (1997). HIV-1: gambling on the evolution of drug resistance. *Nature Medicine* **3**, 268.

Prescott, L. M., Harley, J. P. and Klein, D. A. (1996). *Microbiology,* 3rd edn, William C. Brown, Dubuque IA.

Recchia, G. D. and Hall, R. M. (1995). Gene cassettes: a new class of mobile element. *Microbiol.* **141**, 3015.

Salyers, A. A. and Amábile-Cuevas, C. F. (1997). Why are antibiotic resistance genes so resistant to elimination? *Antimicrob. Agents. Chemother.* **41**, 2321.

Salyers, A. A. *et al.* (1995). Conjugative transposons: an unusual and diverse set of integrated gene transfer elements. *Microbiol. Rev.* **59**, 579.

Tenover, F. C. and Hughes, J. M. (1996). The challenge of emerging infectious diseases: development and spread of multiply-resistant bacterial pathogens. *Science* **275**, 300.

Tomasz, A. and Muñoz, R. (1995). β-Lactam antibiotic resistance in Gram-positive bacterial pathogens of the upper respiratory tract: a brief overview of mechanisms. *Microbiol. Drug Resist.* **1**, 103.

Chapter nine

Biochemical mechanisms of resistance to antimicrobial drugs

Although the individual modes of resistance to antimicrobial drugs are very diverse, they can be grouped into a limited set of general mechanisms that account for most types of resistance encountered in medical practice. These include:

1. conversion of the active drug to an inactive derivative by enzyme(s) synthesized by the resistant cells;
2. loss of sensitivity of the drug target site as a result of:
   (a) covalent modification by enzyme activity in the resistant cells,
   (b) mutation(s) affecting the target, or
   (c) acquisition of genetic information encoding either a drug-resistant form of the target enzyme or overproduction of the drug-sensitive enzyme.
3. Removal of the drug from the cellular interior by drug efflux systems located in the cell envelope.

In addition there are unique modes of resistance not included in this broad classification but which are nevertheless of major medical importance, e.g. vancomycin resistance. The actual level of cellular resistance observed may be due to a combination of factors. In the case of Gram-negative bacteria, for example, resistance often results from a combination of the intrinsically low permeability of the outer membrane together with mechanisms such as drug inactivation or drug efflux.

The rest of this chapter is devoted to examples of the biochemical processes involved in resistance to clinically important drugs used in the treatment of infections caused by bacteria, fungi, viruses and protozoa.

9.1 Enzymic inactivation of drugs

9.1.1 β-Lactams

The destruction of penicillins, cephalosporins and latterly of carbapenems by bacteria that produce β-lactamases is one of the most widespread and most serious forms of microbial resistance. The general inactivation reactions are shown in Figure 9.1. The β-lactam bonds of penicillins and cephalosporins are cleaved to yield the biologically inactive derivatives penicilloic acid and cephalosporanoic acid, respectively. Penicilloic acid is a stable end-product but cephalosporanoic acid spontaneously degrades to several other compounds. As we shall see, the nature of the R-substituent in the amide side chain of β-lactams is important in determining the susceptibility of compounds to β-lactamase attack. The number of β-lactamases and related proteins produced by different bacteria is astonishing – more than 190 such proteins have been identified so far with a wide range of substrate preferences for penicillins and cephalosporins.

Gram-positive β-lactamases

The most important β-lactamase in Gram-positive bacteria is that produced by *Staphylococcus aureus,* which was responsible for the rise in the resistance of this pathogen to penicillin first observed in the late 1940s and 1950s. In many hospitals today more than 90% of the *Staphylococcus aureus* isolates are resistant to the simpler penicillins, due to β-lactamase. The β-lactamase of *Staphylococcus aureus* is an inducible enzyme. Enzyme production is very low in the absence of penicillin or cephalosporin. The addition of minute quantities of antibiotic (as little as 0.0024 μg ml^{-1} of medium) increase enzyme production enormously and the β-lactamase may account for more than 3% of the total protein synthesized by the bacterium. The enzyme, which preferentially attacks penicillins, is released from the bacterial cell and inactivates antibiotic in the surrounding medium. Considerable dilution of β-lactamase occurs and this is the basis of the 'inoculum effect'. A small inoculum of *Staphylococcus aureus* cells may not destroy all the antibiotic in the medium, but the much greater quantity of enzyme produced by a heavy inoculum of cells is able to overcome the challenge. Staphylococcal resistance to penicillin is therefore dependent upon the size of the inoculum.

All the β-lactamases of Gram-positive bacteria are inducible by β-lactams, and the major components of the regulatory system have been defined by studies mainly with *Bacillus licheniformis.* The products of three chromosomal genes, *blaI, blaR1* and *blaR2,* control the expression of the structural

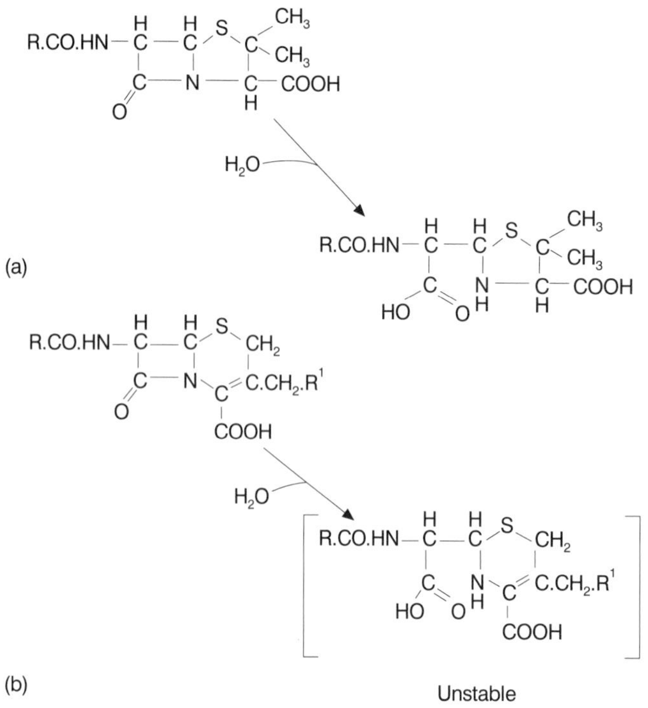

FIGURE 9.1 Inactivation of (a) penicillins and (b) cephalosporins by β-lactamase. While penicilloic acid is relatively stable, the corresponding cephalosporin product is highly unstable and decomposes spontaneously to a complex mixture. R and R^1 represent a wide variety of side chains that may substantially affect the efficiency of β-lactamase attack.

gene for β-lactamase, *blaP*. BlaI, the protein encoded by *blaI*, is a repressor that binds to operator sites between *blaI* and *blaP* to inhibit the transcription of both genes. In the absence of an inducer, BlaI therefore represses the synthesis of β-lactamase. BlaR1 is a large membrane-spanning protein of 601 amino acid residues with a 261 amino acid extracellular domain that interacts with β-lactams in the external environment. The recognition site is located mainly in the region around serine-402 which has close homology with the domains flanking the active-site serine residue of β-lactamase (see below). The binding of an inducing β-lactam is thought to initiate a conformational change in the extracellular domain of BlaR1 which transmits to an intracellular region of the protein via the transmembrane domains. The intracellular region has sequence features of a metallo-peptidase and the conformational changes induced by penicillin binding may activate a latent peptidase activity which would degrade the repressor protein BlaI and permit expression of the structural *blaP* gene.

Although this model provides a possible mechanism for the induction of β-lactamase, it fails to include a role for the product of the *blaR2* gene. The function of the BlaR2 protein is presently unknown but is nevertheless essential for the regulation of β-lactamase synthesis. Clearly, more research is required before the picture of β-lactamase induction in *Bacillus licheniformis* is complete. In *Staphylococcus aureus*, genes encoding the synthesis and regulation of β-lactamase are usually found on plasmids.

Gram-negative β-lactamases

The complex outer envelope of Gram-negative cells makes them intrinsically less sensitive to many β-lactams. However, soon after the introduction of penicillin derivatives, such as ampicillin, with good activity against Gram-negative bacteria, β-lactamase-mediated resistance amongst these pathogens also began to emerge.

Many of the β-lactamases of Gram-negative bacteria are not inducible and are expressed constitutively. However, in a few Gram-negative bacteria, including *Enterobacter cloacae, Citrobacter freundii* and *Pseudomonas aeruginosa,* the enzyme is inducible although the mechanism of induction is quite different from that in Gram-positive bacteria. The inducible β-lactamase genes of Gram-negative bacteria are exclusively chromosomal in location. The structural genes for the enzymes, designated collectively as *ampC,* encode an extended family of sequence-related proteins. Enzyme induction is controlled in *Enterobacter cloacae* and *Citrobacter freundii* by an adjacent *ampR* gene. Unlike the system in Gram-positive bacteria, β-lactams are not directly involved in the induction process. Instead, it is thought that inhibition of the cross-linking of peptidoglycans by β-lactams (Chapter 2) causes an increase in the normal turnover of peptidoglycan in Gram-negative bacteria. The turnover process involves the transport of peptidoglycan fragments generated in the cell wall across the cytoplasmic membrane into the cytoplasm. The transport process is effected by a transmembrane protein encoded by the *ampG* gene that is also essential for the induction of β-lactamase. Another gene, *ampD,* appears to act as a negative regulator of β-lactamase synthesis, since *ampD* mutants produce β-lactamase constitutively. *AmpD* encodes an amidase, AmpD, that specifically hydrolyses the peptidoglycan fragment shown in Figure 9.2. A current suggestion is that the undegraded fragment interacts with the protein encoded by the *ampR* gene referred to above. The AmpR protein normally represses the transcription of the structural gene for β-lactamases. However, the interaction of AmpR with the peptidoglycan fragment facilitates conversion from a repressor to an activating function which 'switches on' expression of β-lactamase. This model implies that the negative regulatory activity of the amidase specified by *ampD* is overwhelmed by the sudden influx of peptidoglycan fragments when the cell is under attack by β-lactams. The Gram-negative induction process is undoubtedly complex and our perception of the details may evolve as further research unfolds.

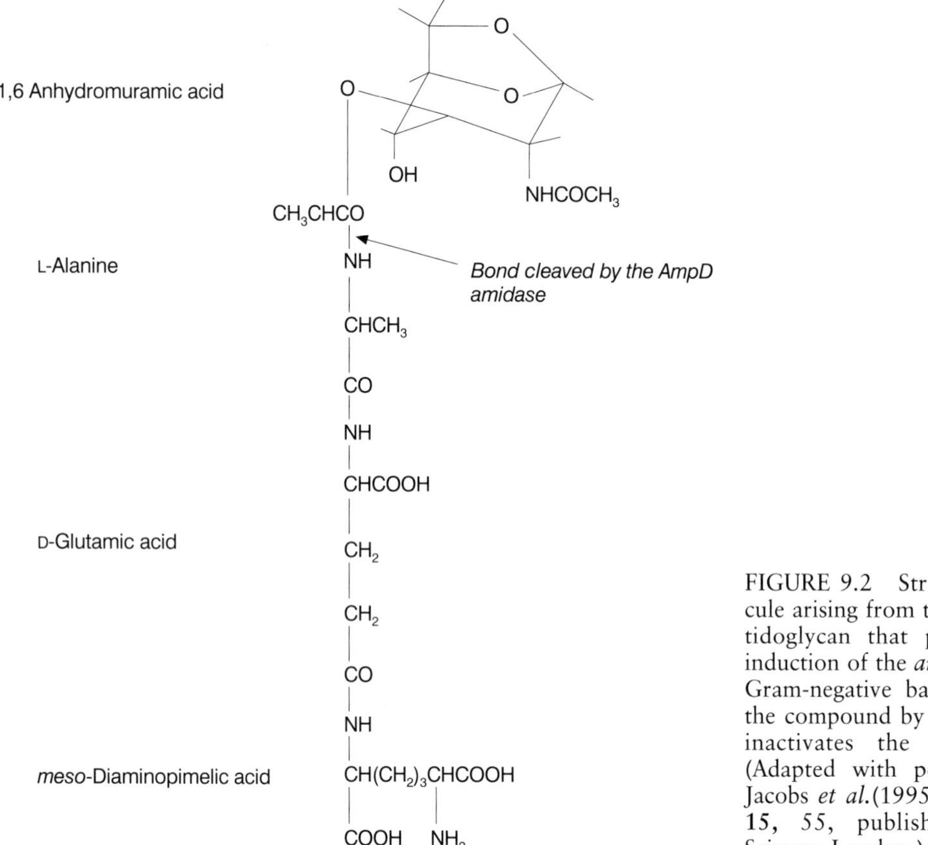

FIGURE 9.2 Structure of the molecule arising from the turnover of peptidoglycan that participates in the induction of the *ampC* β-lactamase of Gram-negative bacteria. Cleavage of the compound by the AmpD amidase inactivates the inducing activity. (Adapted with permission from C. Jacobs *et al.*(1995) *Molec. Microbiol.* **15**, 55, published by Blackwell Science, London.)

Catalytic mechanisms of β-lactamases

Although there is considerable amino acid sequence diversity among β-lactamases, the majority cleave the β-lactam bond by a common catalytic process. These enzymes are known as acyl serine transferases. During the hydrolytic sequence the carbonyl carbon of the β-lactam amide bond transfers to a serine residue at the active centre of the enzyme to form a serine ester-linked β-lactamoyl enzyme complex (Figure 9.3). The β-lactam substrate binds to the active site of the enzyme so as to allow the proton of the γ-hydroxyl group of the serine residue to be abstracted. The resulting activated γ-oxygen atom then attacks the carbonyl group of the β-lactam molecule and the abstracted proton is transferred to the adjacent nitrogen atom. In the next stage a proton is abstracted from a water molecule at the active centre and the activated hydroxyl group attacks the serine–lactamoyl ester bond. The ensuing hydrolysis releases free enzyme and a biologically inactive derivative, either penicilloic or cephalosporanoic acid. It is interesting to compare this reaction (summarized in Figure 9.3) with that between β-lactams and the penicillin-binding proteins (PBPs) involved in peptidoglycan biosynthesis (Chapter 2). In the latter case the β-lactamoyl–PBP complexes, unlike the corresponding complexes in β-lactamases, are relatively resistant to attack by water molecules, resulting in long-lasting inactivation of the PBPs. However, the analogous interactions of the active-site serine residues of β-lactamases and PBPs with β-lactams, and the similarities in the active-site sequences and in the secondary structures of both groups of proteins, suggest that β-lactamases and

PBPs are related, and it is possible that the serine-active-site, β-lactam hydrolysing β-lactamases may have evolved from the same ancestral protein as the present-day PBPs.

Metallo-β-lactamases

Several species of bacteria produce chromosomally mediated β-lactamases that have a zinc ion at the active centre. Until relatively recently these metallo-β-lactamases were considered to be little more than interesting biochemical curiosities. However, their appearance in several troublesome bacterial pathogens, including *Serratia, Bacteroides* and *Aeromonas,* their ability to degrade penicillins, cephalosporins and also the serine-active-site β-lactamase-resistant carbapenems, together with their lack of susceptibility to β-lactamase inhibitors, has highlighted the potential threat of metallo-β-lactamases to the treatment of serious infections.

Hydrolysis of the β-lactam bond by the metallo-β-lactamases does not involve formation of an unstable ester with a serine residue. Instead, the zinc ion at the active centre activates a bound water molecule as a nucleophile. X-ray crystallographic analysis of the metallo-β-lactamase from *Bacillus cereus* suggests that the adjacent aspartate-90 residue also contributes to the activation process by acting as a general base to remove a proton from the water molecule. The abstracted proton is donated to the nitrogen atom of the β-lactam bond and cleavage of the bond ensues. Although most, if not all, of the recognized metallo-β-lactamases are believed to involve a zinc ion in the catalytic process, significant differences in the amino acid sequences and tertiary structures among the enzymes suggest that there may also be differences in the molecular details of the catalytic process.

Approaches to the β-lactamase problem

β-Lactamase-stable compounds

The advent of the semi-synthetic β-lactams during the 1950s offered an escape from the problem of bacterial resistance caused by β-lactamase. Compounds such as methicillin and cloxacillin (Figure 2.14) with bulky substituents in the penicillin side chain were found to be poor substrates for β-lactamase. The affinity of methicillin for staphylococcal β-lactamase is much lower than that of benzyl penicillin, and the maximum rate of hydrolysis of methicillin by this enzyme is only one-thirtieth of that of benzyl penicillin. Until recently methicillin was effective against infections caused by β-lactamase-producing staphylococci, even though its intrinsic antibacterial activity is substantially lower than that of benzyl penicillin. Although it is only slowly degraded by Gram-negative β-lactamases, methicillin is ineffective against Gram-negative infections because its physical characteristics limit its ability to penetrate the outer membrane. To combat the menace of Gram-negative β-lactamases,

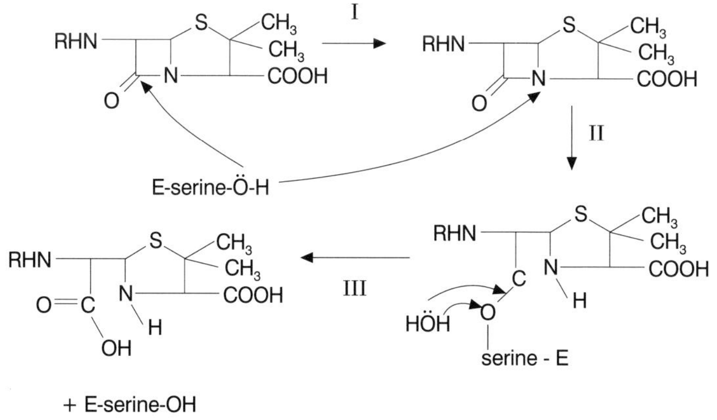

FIGURE 9.3 The essential reactions at the active centre of the serine β-lactamases. I: The proton of the γ-OH of the active-centre serine is abstracted and the resulting activated γ-oxygen atom attacks the β-lactam carbonyl group to form an ester link. The proton is then back donated to the adjacent nitrogen atom. II: Abstraction of a proton from a water molecule at the active centre results in an attack of the activated water OH group on the serine–lactamoyl ester bond to release the degraded β-lactam and regenerated enzyme (III). E represents the rest of the enzyme molecule.

therefore, compounds were needed that both resisted β-lactamase attack and penetrated effectively to the PBPs in the cytoplasmic membrane.

An extensive range of novel β-lactam derivatives has been developed, with good activity against *Escherichia coli* strains producing the most commonly encountered β-lactamase of Gram-negative bacteria, TEM-1. This enzyme, whose name is derived from that of a patient treated for a β-lactam-resistant infection, is encoded on an R-plasmid that transmits readily to other Enterobacteria. The examples in Table 9.1 show that *Escherichia coli* cells producing TEM-1 are effectively inhibited by β-lactamase-resistant compounds. Unfortunately, since the introduction of these antibiotics TEM-1 has undergone mutations near the active centre that markedly increase its ability to hydrolyse several of these valuable agents. Two examples of these mutant enzymes, TEM-12 and TEM-26, are listed in Table 9.1. TEM-26 has two critical amino acid replacements: serine for arginine at position 164 and lysine for glutamate at position 104. These changes dramatically enhance the hydrolytic efficiency of the enzyme against drugs of major importance such as ceftazidime and cefuroxime, and bacteria equipped with this enzyme are markedly more resistant to these drugs (Table 9.1). More than 20 TEM-1-related β-lactamases with increased activity against β-lactams have been discovered so far.

Other bacteria, such as *Enterobacter cloacae*, have evolved a different strategy to combat β-lactamase-resistant antibiotics. As discussed previously, the chromosomal *ampC* gene encoding β-lactamase in this organism is normally an inducible gene that is indirectly regulated via the action of β-lactams on peptidoglycan metabolism. In the modified strategy, the challenge of β-lactamase-stable drugs is countered by mutations in the regulatory cascade that markedly increases the production of enzyme. Large quantities of a β-lactamase with weak activity degrade enough antibiotic to allow the bacteria to survive and proliferate.

The carbapenems, in which the sulphur atom of the β-lactam fused ring system is replaced by a carbon atom (e.g. thienamycin and meropenem in Figure 2.15), generally have excellent stability to the serine-active-site β-lactamases and are highly active against bacteria producing the various forms of TEM (see imipenem in Table 9.1). However, it is now apparent that carbapenems are hydrolysed by the metallo-β-lactamases and also to some extent by some unusual serine-active-site enzymes. Although the latter enzymes are of little clinical significance, the activity of metallo-enzymes against carbapenems poses another serious threat to the control of β-lactam-resistant bacterial pathogens.

Inhibitors of β-lactamases

The discovery of a naturally occurring inhibitor of β-lactamases, clavulanic acid (Figure 2.15), opened the way for valuable synergism with β-lactamase-susceptible drugs such as ampicillin and amoxycillin. Clavulanic acid has little antibiotic activity of its own but is a remarkably effective inhibitor of many β-lactamases of Gram-positive and Gram-negative bacteria. Clavulanic acid and two other clinically useful compounds, sulfazecin (Figure 2.15) and tazobactam, are β-lactams that react irreversibly with the active-site serine of these β-lactamases to form stable, enzymically inactive complexes. However, bacterial evolution is again proving equal to the challenge of inhibitors of β-lactamase. Mutant forms of the enzyme are emerging that are highly resistant to inhibition. Several clavulanic acid-resistant forms of the TEM-1 β-lactamase have been discovered in clinical isolates of *Escherichia coli*, *Klebsiella pneumoniae* and *Proteus mirabilis*. Replacement of methionine at position 68 by the aliphatic amino acids isoleucine, leucine or valine causes a marked increase in the concentration of clavulanic acid required to inhibit the enzyme by 50% (IC_{50}). Other bacterial strains that are resistant to combinations of β-lactams with a β-lactamase inhibitor are characterized by overproduction of β-lactamase. Unfortunately, at present there are no clini-

TABLE 9.1 The effects of TEM β-lactamase and two mutant variants on the hydrolysis of, and bacterial sensitivity to, β-lactams

Enzyme	*Relative rates of hydrolysis (benzyl penicillin = 100)*				*MIC values for* Escherichia coli *(μg ml^{-1})*			
	AMP	CTAX	CTAZ	IMP	AMP	CTAX	CTAZ	IMP
R$^-$	–	–	–	–	4	0.125	0.25	0.25
TEM-1	110	0.07	0.01	<0.01	>256	0.125	0.125	0.25
TEM-12	14	2.4	3.8	>1	>256	0.5	64	0.25
TEM-26	–	7.5	170	–	>256	64	128	0.25

AMP, ampicillin; CTAX, cefotaxime; CTAZ, ceftazidime; IMP, imipenem (a carbapenem). Although no value for the rate of hydrolysis of AMP by TEM-26 is presented, the high minimal inhibitory concentration (MIC) against *Escherichia coli* expressing this enzyme indicates rapid destruction of the antibiotic. R$^-$ indicates bacteria without a resistance plasmid.

cally useful inhibitors of metallo-β-lactamases, although chelating compounds that complex with the essential zinc ion are effective inhibitors in the laboratory.

The three-dimensional structures of both serine-active-site and metallo-β-lactamases have been solved by X-ray crystallography. Two examples are illustrated in Figures 9.4a and b. Hopefully, the information provided by the X-ray data together with synthetic organic chemistry will result in a continuing flow of novel agents to contain the threat of drug-resistant, β-lactamase-producing bacteria.

Classification of β-lactamases

The numbers and diversity of β-lactamases are remarkable; more than 190 bacterial proteins are known that recognize β-lactams as substrates. The classification of this wealth of functionally related enzymes is a major challenge and several systems have been devised. The need for some form of classification first became apparent when enzymes were discovered with differences in their abilities to hydrolyse penicillins and cephalosporins. These differences provided the basis for an early categorization, which was later refined into a scheme with five major groups of enzymes with distinct activities against a range of substrates. Another scheme assigned the enzymes to four major structural types. Details of the most recent and comprehensive classification (the 'Bush' scheme, named after its senior proponent) are available in a review listed at the end of this chapter. In outline, the Bush classification is based upon the functional characteristics of β-lactamases and divides them into three major groups and a fourth, minor group:

Group 1: cephalosporinases that are poorly inhibited by clavulanic acid.

Group 2: penicillinases, cephalosporinases and broad-specificity enzymes that are inhibited by clavulanic acid and other active-site-directed inhibitors of β-lactamases.

Group 3: metallo-enzymes only poorly inhibited by all the conventional β-lactamase inhibitors.

Group 4: a limited number of penicillinases that are not inhibited by clavulanic acid.

Group 2 includes an extended and varied set of enzymes that is further subdivided into eight subclasses according to their substrate and inhibitor profiles. The Bush system of classification is summarized in Table 9.2.

The amino acid sequences of many β-lactamases have been determined either directly or have been inferred from the nucleotide sequences of cloned genes encoding the enzymes. A 'family tree', or dendrogram, constructed from some 88 published sequences reveals a good correlation between the clustering of related sequences and the Bush system of classification. The various analyses of the fascinating diversity of β-lactamases have been of considerable help not only in explaining bacterial resistance to individual antibiotics but also in developing research strategies to combat these

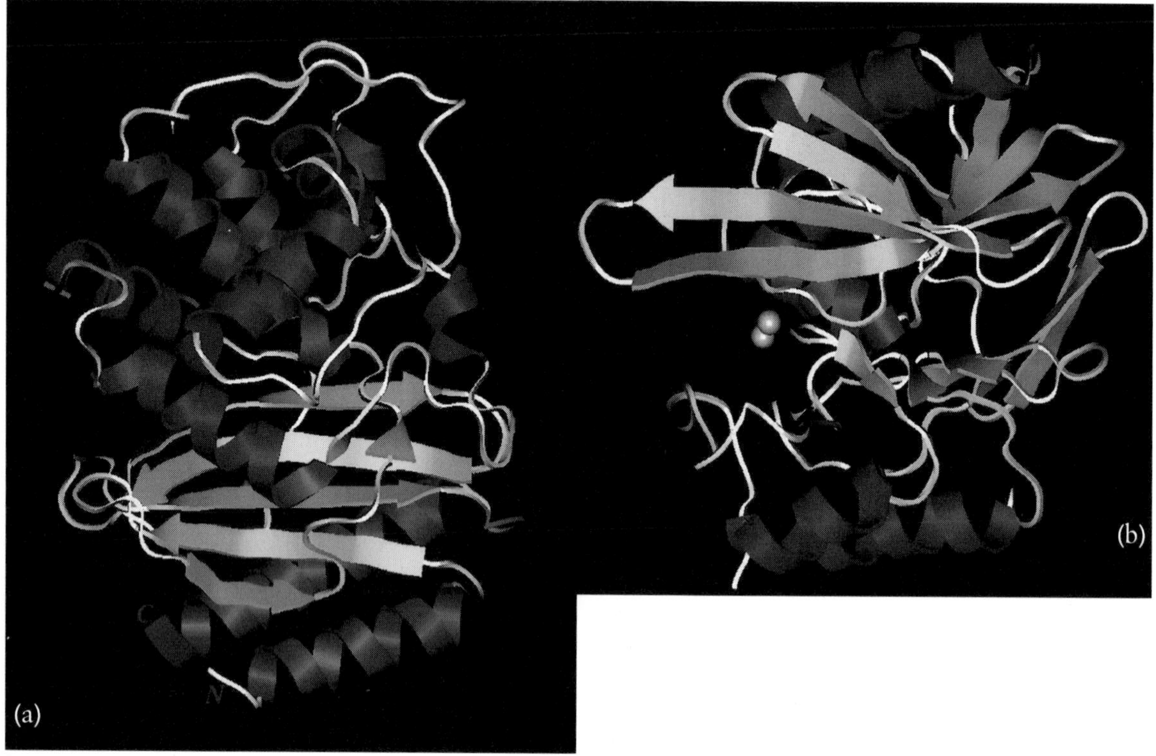

FIGURE 9.4 Three-dimensional structures, as revealed by X-ray crystallographic analysis, of: (a) the TEM-1 serine-active-centre β-lactamase from *Escherichia coli* (reproduced with permission from C. Welsch *et al.* (1993) *Proteins: Structure, Function and Genetics* **16,** 364, published by Wiley-Liss, Inc., a subsidiary of John Wiley & Sons); (b) the zinc β-lactamase from *Bacillus cereus* (reproduced with permission from A. Carfi *et al.* (1995) *EMBO J.* **14,** 4914, published by Oxford University Press). In both diagrams the spiral ribbons represent α-helices; the arrowed ribbons, β-pleated sheets; and the strings, looped regions of the proteins. The zinc atoms at the active centre of the metallo-enzyme are represented by the two silvered spheres. Original data from Protein Data Bank (http://www.pdb.bnl.gov). ID codes 1BTL (TEM-1 serine-active-centre β-lactamase) and 1BME (zinc β-lactamase from *Bacillus cereus*). (*See* facing COLOUR PLATE).

remarkable enzymes. Assuming that novel forms of β-lactamases continue to be discovered, it is likely that further revision of the existing schemes of classification will eventually be needed.

9.1.2 Chloramphenicol

The potential toxicity of chloramphenicol limits its use to the treatment of life-threatening infections, such as typhoid and meningitis, where bacterial resistance to the drug may have serious consequences. Resistance to chloramphenicol is caused mainly by the enzyme chloramphenicol acetyl transferase (CAT) which is widespread among most genera of Gram-positive and Gram-negative bacteria. Genes encoding many variants of the enzyme are known and are either chromosomally or plasmid located. A major subtype in Gram-negative bacteria is found on transposon Tn9. CATs normally exist in solution as trimers with subunit molecular weights of between 24 and 26 kDa.

The catalytic process of chloramphenicol acetyl transferase

CAT metabolizes chloramphenicol in a two-stage process to a 1,3-diacetoxy derivative (Figure 9.5).

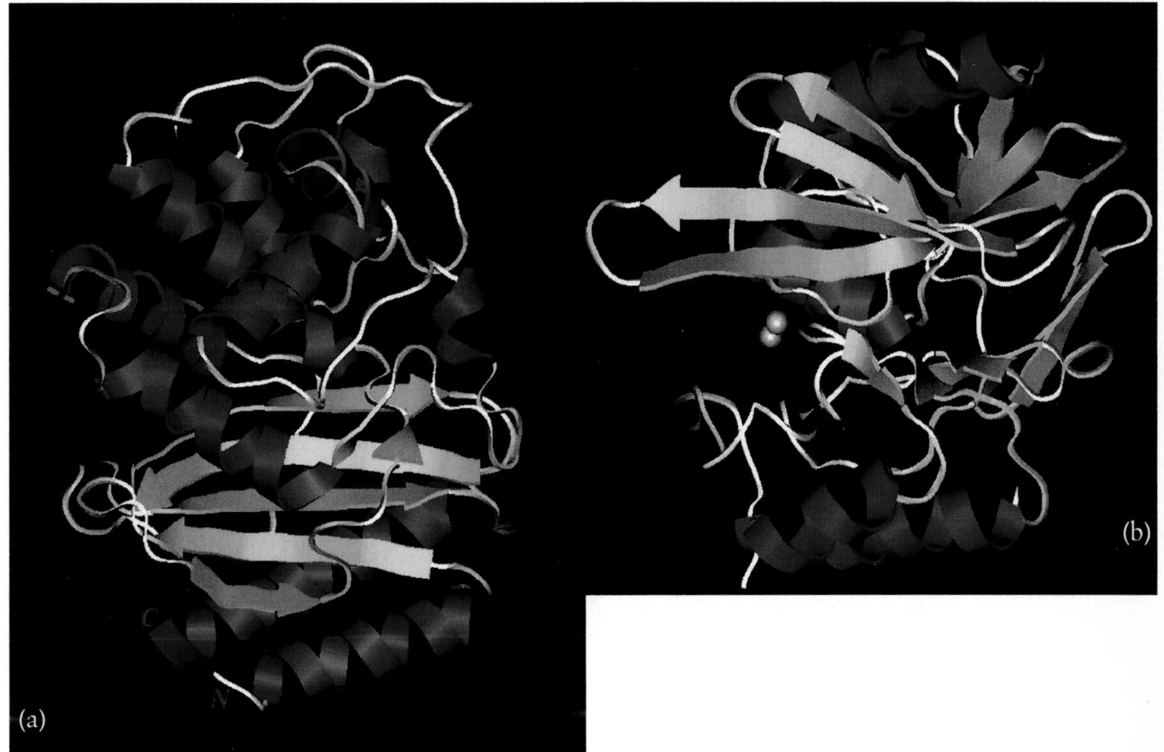

FIGURE 9.4 Three-dimensional structures, as revealed by X-ray crystallographic analysis, of: (a) the TEM-1 serine-active-centre β-lactamase from *Escherichia coli* (reproduced with permission from C. Welsch *et al.* (1993) *Proteins: Structure, Function and Genetics* **16,** 364, published by Wiley-Liss, Inc., a subsidiary of John Wiley & Sons); (b) the zinc β-lactamase from *Bacillus cereus* (reproduced with permission from A. Carfi *et al.* (1995) *EMBO J.* **14,** 4914, published by Oxford University Press). In both diagrams the spiral ribbons represent α-helices; the arrowed ribbons, β-pleated sheets; and the strings, looped regions of the proteins. The zinc atoms at the active centre of the metallo-enzyme are represented by the two silvered spheres. Original data from Protein Data Bank (http://www.pdb.bnl.gov). ID codes 1BTL (TEM-1 serine-active-centre β-lactamase) and 1BME (zinc β-lactamase from *Bacillus cereus*).

TABLE 9.2 The Bush functional classification of β-lactamases *

Functional group	*Preferred substrates*	*Inhibition by: CVA*	*EDTA*	*Some examples*
1	Cephalosporins	–	–	AmpC enzymes of Gram-negative bacteria
2a	Penicillins	+	–	Penicillinases of Gram-positive bacteria
2b	Penicillins, cephalosporins	+	–	TEM-12, TEM-26
2be	Penicillins, cephalosporins, monobactams	+	–	TEM-12, TEM-26
2br	Penicillins, cephalosporins	±	–	TEM-30, TEM-31
2c	Penicillins, carbenicillin	+	–	*Proteus mirabilis* GN79 enzyme
2d	Penicillins, cloxacillin	±	–	*Pseudomonas* C β-lactamases
2e	Cephalosporins	+	–	CepA of *Bacteroides fragilis*
2f	Penicillins, carbapenems	+	–	Sme-1
3	Most β-lactams including carbapenems	–	+	L1 of *Xanthomonas maltophila*
4	Penicillins	–	?	Penicillinase of *Pseudomonas cepacia*

*Adapted from Bush, K. *et al.* (1995) *Antimicrobial Agents and Chemotherapy* 3, 1211. CVA, clavulanic acid; EDTA, ethylenediaminotetraacetic acid, a chelator of divalent metal ions. Further details of enzymes referred to in the table can be obtained from the review quoted above.

In the first stage the antibiotic is converted to the 3-acetoxy compound using acetyl coenzyme A as an essential cofactor. A slow, non-enzymic rearrangement then transfers the acetoxy group to the 1-position. A further round of enzymic acetylation at the 3-position generates the final 1,3-diacetoxy product, although this reaction is much slower than the first step owing to the impaired fit of the 1-acetoxychloramphenicol into the active site of the enzyme. As both the mono- and diacetoxy derivatives are inactive as antibiotics, the two-stage acetylation sequence is biologically inefficient but it is nevertheless a consequence of the spontaneous shift of the acetyl group from the 1- to the 3-position. Although the chloramphenicol molecule is potentially vulnerable to various forms of metabolic inactivation, including dehalogenation, reduction of the nitro group and hydrolysis of the amide bond, the acetylation mechanism is the overwhelmingly significant contributor to resistance among bacterial pathogens.

Of the various forms of CAT defined by their structural and biochemical properties the type III enzyme (CAT_{III}) has been most thoroughly studied. Solution of the structure of CAT_{III} by X-ray crystallography was a major contributor to the understanding of the catalytic mechanism (Figure 9.6). The trimeric holoenzyme has three identical active sites in the interfacial clefts between the monomers. Histidine-195 in one face of each cleft acts as a general base to abstract a proton from the 3-hydroxyl group of chloramphenicol. The resulting oxyanion attacks the 2-carbonyl carbon atom of acetyl coenzyme A to yield a tetrahedral intermediate. The oxygen atom of the intermediate is hydrogen bonded to the hydroxyl group of serine-148. The intermediate may also form hydrogen bonds with a water molecule linked to threonine-174. Finally, the tetrahedral intermediate collapses to yield 3-*O*-acetylchloramphenicol and free coenzyme A. The first-stage acetylation by CAT_{III} is an extremely efficient reaction with a turnover rate of 600 molecules s^{-1}.

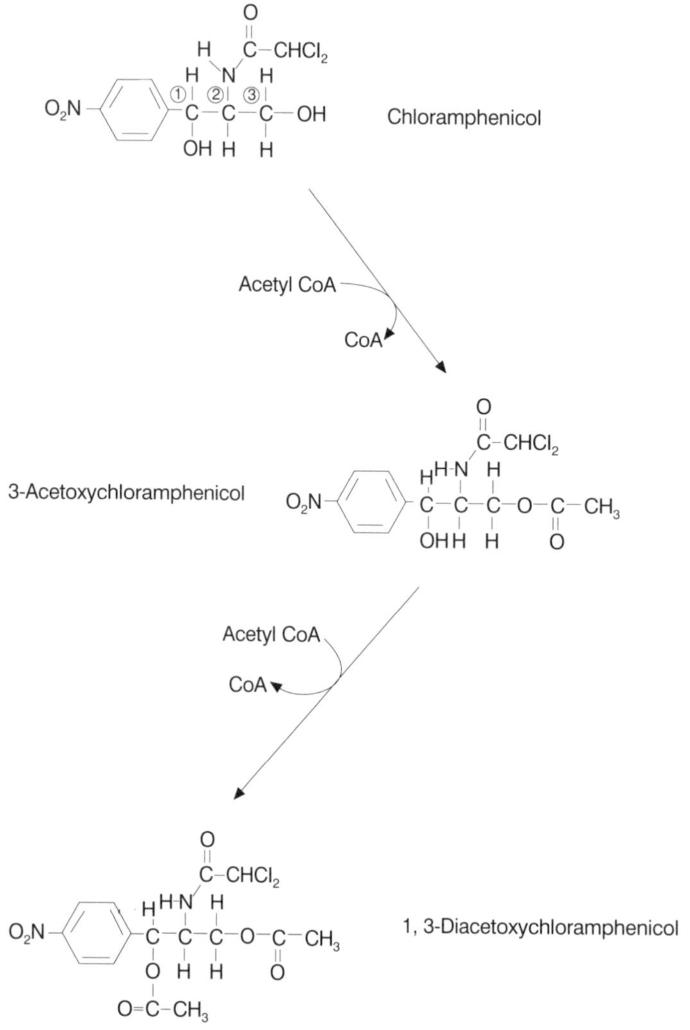

Chloramphenicol

Acetyl CoA

CoA

3-Acetoxychloramphenicol

Acetyl CoA

CoA

1, 3-Diacetoxychloramphenicol

FIGURE 9.5 Inactivation of chloramphenicol by chloramphenicol acetyl transferase (CAT). 3-Acetoxychloramphenicol is formed first, followed by a non-enzymic shift of the acetyl group to the 1-position. A second enzymically catalysed acetylation at the 3-position yields the 1,3-diacetoxy derivative.

Physiology of chloramphenicol acetyl transferase synthesis

Gram-positive bacteria

The genes encoding CATs in Gram-positive bacteria such as *Staphylococcus* spp. and *Bacillus* spp. are inducible by chloramphenicol. However, the mechanism of induction does not involve increased transcriptional activity but an activation of the translation of the mRNA for the enzyme. Investigation of the inducible *cat* gene from *Bacillus pumilus* revealed that there is an 86 bp (base pair) region immediately 5′ to the coding sequence for the enzyme. Within the 86 bp region there are two distinct domains: domain A contains a ribosome-binding site (RBS-2), a translation initiation codon (GTG) and an open reading frame of nine codons ending in an upstream inverted-repeat sequence within domain B. The latter domain has two 14 bp inverted repeat sequences separated by 12 bp. The downstream inverted repeat spans the specific ribosome-binding site (RBS-3) for the *cat* gene transcript.

The sequence of domain B suggests that its transcript contains a stable stem–loop structure that may negatively regulate translation of the *cat*

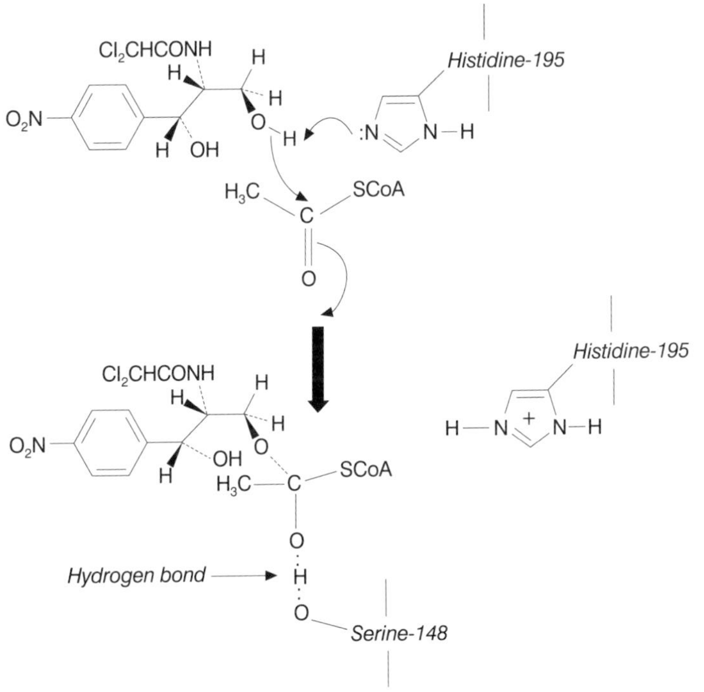

FIGURE 9.6 The key events in the catalytic mechanism of chloramphenicol acetyl transferase. Histidine-195 at each of the three active centres acts as a general base to abstract a proton from the 3-OH group of the antibiotic. The resulting activated oxygen attacks the 2-carbonyl group of acetyl coenzyme A to generate a tetrahedral intermediate that is hydrogen bonded to the γ-OH group of serine-148 of the enzyme. Subsequently the intermediate collapses to release 3-acetoxychloramphenicol, coenzyme A and free enzyme. (Reaction mechanism adapted with kind permission of I. A. Murray and W. V. Shaw and the American Society of Microbiology: *Antimicrob. Agents Chemother.* **41**, 1 (1997).)

gene transcripts. A detailed analysis of the transcriptional mechanism has shown that while the *cat* coding sequence and the upstream 86 bp region can be transcribed into a single mRNA molecule, about 50% of the observed transcripts terminate immediately after the regulatory region. This suggests that the mRNA stem–loop also acts as a weak transcription termination signal. Furthermore, the full-length transcripts are not efficiently translated into enzyme protein in the absence of chloramphenicol because RBS-3 is hidden within the stem–loop structure. Addition of the antibiotic causes a ribosome engaged in translating the leading sequence of domain A to stall (remember that chloramphenicol inhibits protein synthesis, Chapter 5). The stalled ribosome masks sequences in the mRNA, leading to destabilization of the inhibitory stem–loop and RBS-3 is made available to initiate translation. Remarkably, therefore, the ability of chloramphenicol to inhibit protein biosynthesis is exploited to facilitate the efficient translation of the mRNA and synthesis of the enzyme that inactivates the antibiotic. The destabilization of the stem–loop has only a small effect in relieving transcription termination so that the induction mechanism is largely dependent on increasing the efficiency of translation. This mechanism of induction is referred to as translational attenuation, and we shall see it in action again in inducible resistance to erythromycin and tetracyclines.

Gram-negative bacteria

CAT synthesis is constitutive in most Gram-negative bacteria. However, CAT synthesis in *Escherichia coli* is subject to catabolite repression. Synthesis is faster in cultures grown on glycerol compared with glucose-supported cultures. Cyclic AMP complexed with the catabolite activator protein (CAP) is required for optimal CAT synthesis. During glucose-supported growth the intracellular levels of cyclic AMP are low and the uncomplexed

CAP cannot bind to the promoter region that regulates the *cat* gene. Transcription of the *cat* gene is therefore inefficient. Conversely when cyclic AMP levels rise the CAP–cyclic AMP complex binds to the promoter region, facilitating the interaction of DNA-dependent RNA polymerase with the structural gene and its transcription.

Xenobiotic acetyl transferases (XATs): a new class of chloramphenicol-inactivating enzymes

The extended family of CATs is related by a range of structural and mechanistic characteristics and their specificity for chloramphenicol. Recent cloning studies are bringing to light genes encoding acetyl transferases which acetylate not only chloramphenicol but also other substrates, including antibiotics of the streptogramin group. However, the level of bacterial resistance conferred by the XATs, which have no homology to the known CATs, appears to be quite modest and their contribution to the problem of clinical resistance is uncertain.

9.1.3 Aminoglycosides

Bacterial resistance to aminoglycosides can result from mutations affecting the ribosomes and also from changes in cellular permeability, but the most important cause of resistance is due to enzymically catalysed modifications of the antibiotics. Although there is an extensive range of aminoglycoside-modifying enzymes, only three major types of enzyme-catalysed reactions are known:

1. *N*-acetylation of vulnerable amino groups using acetyl coenzyme A as the acetyl donor;
2. *O*-adenylylation involving the transfer of an AMP residue from ATP to certain hydroxyl groups; and
3. *O*-phosphorylation of hydroxyl groups with ATP acting as the phosphate donor.

Typical reactions involving streptomycin and kanamycin A are shown in Figure 9.7. Streptomycin is subject to both adenylylation and phosphorylation but it is not a substrate for the *N*-acetyl transferases. The *N*-acetyl transferases are specific for the amino groups of other aminoglycosides in four different positions (1-, 3-, 6′- and 2′-). The *O*-adenylyl transferases attack OH groups in 2″, 3″ and 4-positions and the *O*-phosphoryl transferases target OH groups in the 3′-, 3″ and 4-positions. Isozymes exist for many of these enzymes, encoded by at least 30 different genes. A bifunctional enzyme that catalyses both *N*-acetyl and *O*-phosphoryl transferase activities appears to be the result of gene fusion, with protein domains responsible for each enzymic activity being derived from different genes. For a comprehensive account of the many enzymes that metabolize aminoglycosides the reader is referred to a review listed at the end of this chapter. Resistance to aminoglycosides caused by antibiotic metabolism is widespread among both Gram-positive and Gram-negative bacteria and the dissemination of the genes responsible is helped by their association with transposons and transmissible plasmids.

The exact cellular location of the aminoglycoside-modifying enzymes is somewhat uncertain. Because the target of aminoglycoside action is at the ribosome, the cytoplasm was thought to be the most likely location for the inactivating enzymes. However, a cytoplasmic location would be expected to be relatively inefficient in protecting the ribosomes because the enzymes are only synthesized in small amounts. It would make more ‘sense’, therefore, if the enzymes were secreted into the periplasmic space of Gram-negative bacteria where aminoglycoside inactivation would occur before entering the cytoplasm. However, it not clear how the essential co-substrates ATP and acetyl coenzyme A could gain access to the periplasm. Nevertheless, there is evidence for the location of at least one *O*-adenylyl transferase in the periplasmic space. Signal sequences of 20–30 amino acids at the *N*-termini of proteins ensure the export of proteins across the cytoplasmic membrane into the periplasm. Many members of the *N*-acetyl transferase family have these sequences, although they are not found among the

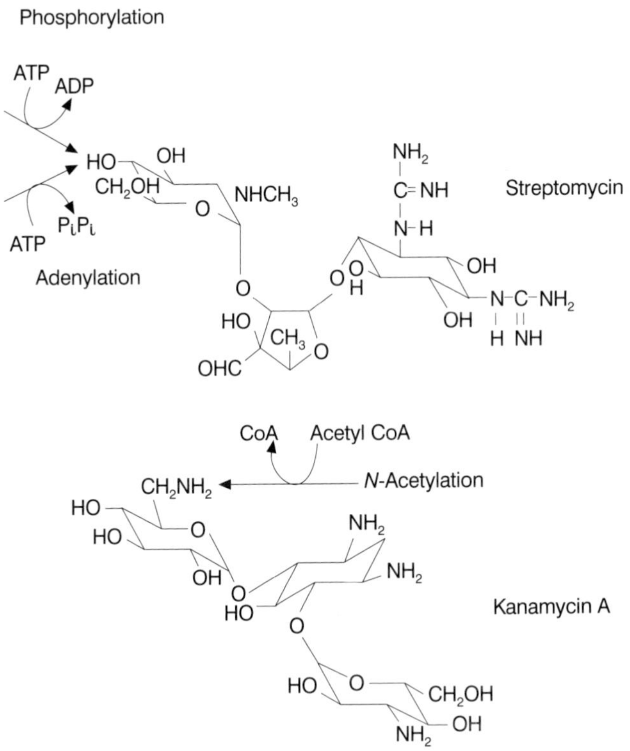

Phosphorylation

ATP ADP

HO OH CH2OH NHCH3 Streptomycin

ATP PiPi

Adenylation

CoA Acetyl CoA

CH2NH2 N-Acetylation

Kanamycin A

FIGURE 9.7 Three modes of enzymic inactivation of aminoglycoside antibiotics. Unlike kanamycin A, streptomycin is not subject to *N*-acetylation, while kanamycin A is also inactivated by O-adenylylation and O-phosphorylation. An extensive array of bacterial enzymes is involved in the inactivation of aminoglycosides. A review listed at the end of the chapter provides further details.

O-phosphoryl transferases. Thus although the periplasmic space appears to be an optimum location for aminoglycoside-modifying enzymes, this may not always be the case.

In most bacteria the expression of genes encoding aminoglycoside-modifying enzymes is not subject to regulation. However, the genes for *N*-acetyl transferases in *Serratia marcescens* and *Providencia stuarti* appear to be tightly regulated, although in some clinical isolates of aminoglycoside-resistant strains the control of expression is relaxed.

Surveys of the mechanisms of aminoglycoside resistance among bacterial pathogens from countries around the world reveal a complex pattern in which many organisms harbour a combination of several genes, either chromosomally or plasmid-located, expressing different modes of aminoglycoside metabolism. Resistance to a range of aminoglycosides is especially marked among *Citrobacter, Enterobacter* and *Klebsiella* species, although other Gram-negative bacteria also exhibit resistance. Successive aminoglycosides, both naturally occurring and semi-synthetic, have been able to combat resistant pathogens. However, this strategy has been only partly successful in meeting the challenge of the many forms of aminoglycoside-inactivating enzymes. Enzymic inhibitors comparable with the β-lactamase inhibitors have been hard to come by and none has achieved clinical application.

9.2 Modification of drug targets

9.2.1 β-Lactams

While the most common mechanism of resistance to β-lactam antibiotics is that of inactivation by β-lactamases, resistance can result from amino acid changes in penicillin-binding proteins (PBPs) that depress their affinity for β-lactams. Mosaic genes encoding hybrid PBPs with reduced affinity for β-lactams were described in Chapter 8. Such genes confer resistance to both penicillins and cephalosporins in meningococci, gonococci and streptococci. In this section we shall concentrate on another example of β-lactam resistance caused by a target change that is causing grave concern: the resistance of *Staphylococcus aureus* to methicillin.

This organism was rightly regarded as a particularly dangerous pathogen in the pre-antibiotic era. The introduction of benzyl penicillin created a fortunate interlude during which staphylococcal infections responded readily to the new drug. The subsequent rise of β-lactamase-mediated resistance in staphylococci was initially countered by the introduction by the semi-synthetic, β-lactamase-stable, methicillin. However, the 1980s saw the emergence of methicillin-resistant *Staphylococcus aureus* (MRSA) which is also resistant to all other β-lactams. Resistance is caused by the acquisition of a transposon-located novel gene, *mecA,* that encodes a novel PBP (PBP2′, otherwise designated as PBP2a) with very low affinity for all β-lactams. The transposon integrates into the chromosome of *Staphylococcus aureus* and is distributed both within and between staphylococcal species. Remarkably, the PBP2′ protein takes over the function of all the other PBPs, thus rendering the growth of *mecA*$^+$ bacteria resistant to methicillin and other β-lactams.

Some bacterial strains have an upstream regulatory region, *mecR1–mecI,* that negatively controls the expression of *mecA*. In the absence of methicillin, synthesis of the MecR1 protein is repressed. Resistance is slowly induced to low levels by methicillin. Mutations in, or complete loss of, the *mecI* region permit the synthesis of both MecR1 and PBP2′, giving rise to high levels of methicillin resistance. There is marked sequence homology of *mecR1* and *mecI* with the *blaR1* and *blaI* genes that regulate β-lactamase synthesis in Gram-positive bacteria. Consequently the *mecA* gene is also regulated by *blaR1* and *blaI* and PBP2′ expression is induced by β-lactams that are recognized by the BlaR1 protein.

Although PBP2′ replaces the function of the other PBPs in *Staphylococcus aureus,* its precise function in peptidoglycan synthesis is uncertain and its activity leads to peptidoglycan with a lower than normal degree of cross-linking. The level of PBP2′ synthesis in *mecA*$^+$ bacteria does not correlate closely with the level of resistance, and it is known that other genes contribute to methicillin resistance, possibly even including genes that encode 'super' penicillinases capable of degrading methicillin.

9.2.2 Erythromycin

Although the precise details are unknown, the inhibition of protein synthesis by erythromycin on ribosomes depends principally upon its interaction with 23S rRNA (Chapter 5). Bacterial strains exist that enzymically inactivate erythromycin and other macrolide antibiotics. However, macrolide metabolism does not contribute significantly to the problem of clinical resistance. Most macrolide-resistant Gram-positive pathogens, including *Staphylococcus aureus* and *Streptococcus* spp., harbour plasmid-borne genes (*erm*) that encode a family of *N*-methyl transferases, or methylases as they are also known. The substrates for these enzymes are specific adenine residues in 23S rRNA in the peptidyl transferase circle or domain involved in the interaction of erythromycin with the ribosome. There are no N^6-methylated adenine residues in the 23S rRNA of wild-type, erythromycin-sensitive bacteria. In contrast, N^6-dimethylated adenine appears in the 23S rRNA of resistant *ermA*$^+$ *Staphylococcus aureus* cells grown in medium containing erythromycin. Evidence that the modified 23S rRNA confers ribosomal resistance is provided by the observation that 70S ribosomes reconstituted with 23S rRNA from resistant *Bacillus subtilis* and ribosomal protein from erythromycin-susceptible cells are resistant to erythromycin. The critical adenine residue targetted by the *erm*-mediated methylase in *Escherichia coli* is A-2058 in the peptidyl transferase domain.

Methylation of A-2058 probably disrupts the tertiary structure of this domain and prevents its inhibitory interaction with erythromycin.

The *N*-methylase products of the *erm* gene family use *S*-adenosylmethione as the methyl donor. Some of the enzymes transfer a single methyl group to the N^6-position of adenine, whereas others donate two groups. Dimethylases may confer a higher level of resistance to a broader range of macrolide antibiotics compared with the monomethylases. The 23S rRNA substrate is presented to the methylases as a component of nascent ribosomes rather than as part of the mature, functioning particles.

Regulation of *erm* gene expression

The expression of *erm* genes is inducible by erythromycin. The inducing activity of erythromycin is closely linked to its ability to inhibit ribosomal function, and erythromycin derivatives devoid of inhibitory activity cannot induce expression of the *N*-methylases. The mechanism of erythromycin induction of the *N*-methylases is analogous to that of chloramphenicol acetyl transferase by chloramphenicol, i.e. induction is achieved by translational attenuation rather than by increased gene transcription. Translation of the *erm* mRNA is slow and inefficient in the absence of erythromycin because of the unfavourable conformation of a 141-nucleotide leader sequence upstream of the open reading frame (orf) for the enzyme. The secondary structure of the leader sequence in some way masks the first two codons of the orf as well as the ribosomal binding site for the mRNA. Translation of the leader sequence is inhibited when erythromycin binds to the ribosome. Paradoxically this increases the efficiency of subsequent translation, probably because the secondary structure of the leader sequence undergoes a conformational rearrangement to a new state consistent with a more rapid read-through of the orf and increased enzyme synthesis. The detailed mechanism by which inhibition of translation of the leader sequence facilitates the favourable conformational change remains to be defined.

9.2.3 Quinolones

The role of the quinolones in the treatment of bacterial infections has grown steadily with the introduction of novel derivatives with broader spectra of action than the progenitor compound, nalidixic acid. The emergence of resistance accompanying this increasing use of quinolones is due largely to mutations in the A subunit of the target enzyme, topoisomerase II, or DNA gyrase (Chapter 4), although drug efflux may also contribute to the problem in Gram-negative bacteria. Enzymic inactivation of quinolones has not so far been detected as a mechanism of bacterial resistance.

The A subunits of the tetrameric enzyme (A_2B_2) are responsible for introducing double-stranded breaks in DNA and for subsequent resealing of the breaks during the negative supercoiling process. The region around serine-83 of the A subunit is critical for the enzyme–quinolone interaction and resistant bacteria carry several mutations between residues 67 and 106. Most of these involve substitution of serine-83 by non-polar, more bulky amino acids, such as leucine, alanine or tryptophan. In some bacteria, including *Campylobacter jejuni*, *Klebsiella pneumoniae* and *Pseudomonas aeruginosa*, threonine is the normal amino acid at position 83 instead of serine and these organisms are intrinsically about tenfold more resistant to quinolones. The greater bulk of threonine at position 83 probably hinders the optimal binding of quinolone to the target site.

9.2.4 Streptomycin

In addition to inactivation by aminoglycoside-modifying enzymes, resistance to streptomycin also arises from mutations affecting the target site in the 30S ribosomal subunit (Chapter 5). Streptomycin still finds some application in the treatment of tuberculosis. DNA analysis of clinical isolates of resistant *Mycobacterium tuberculosis* shows that about 70% have mutations affecting the *rpsL* gene that codes for protein S12 which plays a significant, though undefined role, in the binding of streptomycin to the ribosomal subunit. The mutations cause replacement of lysine-43 or lysine-88 by arginine and a consequent loss of binding of streptomycin to the ribosome. Mutations affecting the highly conserved position 904 of 16S rRNA also cause streptomycin resistance in some *Mycobacterium tuberculosis* isolates. Ribosomal resistance to streptomycin is found in clinical isolates of other bacteria, including *Neisseria gonorrhoeae, Staphylococcus aureus* and *Streptococcus faecalis*.

9.2.5 Rifampicin

Resistance to rifampicin is proving to be a considerable threat to the successful treatment of tuberculosis. Originally this drug was highly effective against *Mycobacterium tuberculosis* but mutations affecting the β subunit of the target enzyme, DNA-dependent RNA polymerase (Chapter 4) are responsible for the loss of bacterial sensitivity to rifampicin. Most resistant clinical isolates of *Mycobacterium tuberculosis* have replacements at serine-531 or histidine-526. These changes, caused by mutations near the centre of the *rpoB* gene which encodes the β subunit, are readily detected by DNA analysis which provides a rapid test for drug resistance. Mutations conferring rifampicin resistance do not affect the growth of *Mycobacterium tuberculosis*, whereas rifampicin resistance in *Escherichia coli* causes slower growth. The reason for this difference is unknown but it may be associated with the naturally slower growth of *Mycobacterium tuberculosis* compared with that of *Escherichia coli*.

9.2.6 Inhibitors of dihydrofolate reductase (DHFR)

Trimethoprim

The basis of one major form of resistance to trimethoprim is analogous to that of methicillin resistance, namely the resistant cells acquire additional genetic information for an enzyme with reduced susceptibility to the drug. In Gram-negative bacteria 16 different genes for trimethoprim-resistant DHFRs have been identified. Mostly these genes are plasmid-borne, although they may temporarily reside on the chromosome because of their association with transposons. The enzymes fall into two families. Family 1 has five members with polypeptide chains sharing 64–88% sequence identity. The enzymes of family 1 are homodimeric proteins with IC_{50} values between 1 μM and 100 μM, compared with an IC_{50} for the wild-type DHFR of 1 nM. Family 2 is larger, with 11 members more closely related than those of family 1, with sequence identities of 78–86%. The enzymes of family 2 are all homotetramers and are highly resistant to trimethoprim, with IC_{50} values greater than 1 mM. The most widely distributed trimethoprim-resistant DHFR amongst Gram-negative bacteria is encoded by the *dhfrI* gene and belongs to family 1. The *dhfrI* gene is located within a highly mobile cassette associated with a promiscuous transposon (Tn7) that inserts into the chromosomes of many bacteria.

Some trimethoprim-resistant bacteria overproduce modified DHFRs. For example, a highly resistant strain of *Escherichia coli* overproduces by about 100-fold an enzyme that is about threefold more resistant to trimethoprim than the wild-type enzyme. Trimethoprim-resistant isolates of *Haemophilus influenzae* (an important clinical target) also overproduce DHFR, although in this case the enzyme is 100- to 300-fold more resistant to the drug. Enzyme overproduction in both species of bacteria is associated with mutations in the promoter sequences that control expression of the structural genes.

Interestingly, the kinetic parameters of trimethoprim-resistant enzymes for the normal substrates, dihydrofolate and NADPH, are essentially unchanged. The mutant enzymes are therefore quite competent to take over the metabolic functions of the drug-sensitive enzyme.

Sulphonamides

Trimethoprim is usually administered in combination with a sulphonamide such as sulphamethoxazole. Unfortunately, bacterial resistance to sulphonamides often coexists with trimethoprim resistance. Sulphonamide resistance is often due either to mutations in the chromosomal gene, *dhps*, that mediates dihydropteroate synthase, or to the acquisition of plasmid-borne genes coding for sulphonamide-resistant forms of the enzyme. In *Neisseria meningitidis* there are two variants of a chromosomally mediated resistant dihydropteroate synthase, one of which may be the result of recombination between two related *dhps* genes. At least two types of plasmid-borne *dhps* genes code for sulphonamide-resistant enzymes in Gram-negative bacteria. In all of these resistant enzymes the Michaelis constants (K_m) for the natural substrate, *p*-aminobenzoic acid, are similar to that of the drug-sensitive enzyme.

Pyrimethamine and cycloguanil

Inhibitors of DHFR, including pyrimethamine and the liver metabolite of proguanil, cycloguanil (Chapter 6), have been central to the prophylaxis and treatment of malaria for more than 50 years. The relentless increase in resistance to these drugs in many parts of the world is therefore a major threat to the containment of one of the most prevalent infections on Earth. Because of the many technical difficulties in working with protozoal parasites, the definition of the mechanisms of resistance in naturally occurring infections has been extremely difficult. Much of the available biochemical information has therefore been obtained with drug-resistant malarial protozoa developed in the laboratory. Nevertheless it is believed that this information gives a reasonable indication of the nature of drug resistance in the 'field'. Furthermore, by using technologies such as the polymerase chain reaction for genetic analysis it should soon be possible to correlate laboratory and field studies of drug-resistant protozoa.

Resistance to pyrimethamine in the most dangerous malarial parasite, *Plasmodium falciparum*, is commonly due to a single-point mutation in DHFR that changes serine-108 to asparagine. This mutation has been detected in isolates from the laboratory and from patients but does not confer cross-resistance to cycloguanil. However, when serine-108 is replaced by threonine together with an alanine-to-valine replacement at position 16, the parasite becomes resistant to cycloguanil but not to pyrimethamine. Strains resistant to both drugs have an asparagine at position 108 in addition to several other point mutations. The various mutations associated with DHFR resistance appear to be involved with the active site of the enzyme or affect those amino acids known to be concerned with the binding of inhibitors to susceptible forms of the enzyme.

Certain strains of *Plasmodium falciparum* and also of another protozoal pathogen, *Leishmania major*, resist pyrimethamine by overproducing DHFR, either by gene duplication or by increasing expression levels.

9.2.7 Inhibitors of HIV reverse transcriptase

The introduction of AZT (3′-azido-3′-deoxythymidine, Chapter 4) was a therapeutic landmark in the treatment of AIDS. AZT has been followed by other nucleoside inhibitors of the reverse transcriptase (RT) of HIV and by non-nucleoside inhibitors such as nevirapine. Unfortunately, the remarkable propensity of HIV for mutation (Chapter 8), combined with the inability of RT inhibitors to suppress viral replication by more than 90%, leads to the rapid emergence of drug-resistant strains of the virus. Nucleotide sequence analysis of resistant viruses isolated from AIDS patients treated with AZT reveals successive amino acid changes near the active site of RT as the level of viral resistance increases. However, the susceptibility of the enzyme isolated from the resistant viruses to the triphosphate of AZT (the intracellular inhibitor) is similar to that of the wild type irrespective of the level of viral resistance to the drug. A full explanation for the nature of HIV resistance to AZT is therefore elusive. Could it be that the triphosphate of AZT is not the ultimate inhibitor of RT in virus-infected cells, or might the mutant enzymes adopt different, inhibitor-resistant conformations in the cells? Further research is needed in order to answer these questions.

The non-nucleoside compound, nevirapine, inhibits RT directly without the need for intracellular metabolism. Unfortunately, nevirapine therapy also leads to the development of drug-resistant HIV. However, in this case mutations in RT are demonstrably associated with enzyme resistance to the drug. X-ray analysis of a susceptible form of the enzyme complexed with nevirapine shows that the binding pocket for the drug consists of two β-strands of amino acids 100–110 and 180–190. The mutations that confer enzyme resistance to nevirapine lie among the residues within these domains. Nevirapine is equally inhibitory to both nucleoside-sensitive and nucleoside-resistant strains of HIV.

9.2.8 Inhibitors of HIV protease

These drugs (Chapter 6) form part of the standard triple therapy for AIDS in combination with two different inhibitors of RT. Triple therapy is considered essential because mutant viruses with drug-resistant proteases emerge readily when protease inhibitors are given as a single therapy. The development of resistance during monotherapy with the protease inhibitor, ritonavir (Figure 6.3), has been documented in considerable detail. The gradual loss of effectiveness of the drug in AIDS patients is associated with a sequential accumulation of mutations in the target enzyme. The rate at which mutations appear is inversely related to the concentration of drug in the patient's blood, strongly suggesting that blood levels of the drug should be maintained as high as possible in order to minimize viral replication. Mutations at single loci in HIV protease are associated with low-level viral resistance to ritonavir: a seven- to ten-fold increase in resistance requires the accumulation of three to four mutations, and high-level resistance (> 20-fold) of four to five mutations. Of the nine amino acid changes contributing significantly to resistance, replacement of valine-82 is probably the most important. X-ray analysis of the susceptible enzyme shows that this residue interacts directly with ritonavir. Ritonavir-resistant viruses are only partially cross-resistant to other protease inhibitors, indicating that there are significant and clinically important differences in the molecular interactions of the various inhibitors with the mutant enzymes.

9.2.9 Acyclovir

Infections caused by the various herpes viruses range from the relatively trivial to severely disabling or even life-threatening. One the most important applications of acyclovir (Figure 4.6) is in the treatment of herpes infections in immunosuppressed patients, in whom drug-resistant forms of the viruses develop most readily. Acyclovir is a pro-drug that is first converted to the monophosphate derivative by thymidine kinase (TK) encoded in the viral genome, and subsequently to the inhibitory acyclovir triphosphate by enzymes of the infected host cell. One relatively uncommon form of resistance to acyclovir results from mutations affecting the viral DNA polymerase which is the ultimate target for the drug. The modified enzyme has a diminished affinity for acyclovir triphosphate while retaining a relatively unchanged ability to bind the four nucleoside triphosphates required for viral DNA synthesis. However, by far the most common mechanism of resistance to acyclovir in immunosuppressed patients depends on mutations to the viral TK. The acquisition of an inappropriately placed stop codon results in a truncated polypeptide with little or no enzymic activity. Viruses lacking TK activity cannot phosphorylate acyclovir and also lose the characteristic ability of herpes viruses to reactivate from a dormant state in neuronal cells. Mis-sense mutations in viral TK, on the other hand, allow the enzyme to retain its affinity for thymidine while markedly depressing that for acyclovir and, furthermore, do not affect the reactivational capability of the virus.

9.3 Drug efflux pumps

The protective ability of living cells to pump toxic chemicals out of the cytoplasm into the external environment occurs in many organisms, ranging from bacteria to mammals. The pumps, more formally known as efflux proteins, reside in the cytoplasmic membrane. In the case of Gram-negative bacteria these cytoplasmic membrane proteins are functionally linked to proteins that bridge both the periplasmic space and outer membrane. The term 'pump' is descriptive since it implies that energy is needed to fulfil its function and, indeed, the drug efflux systems expend energy in transporting compounds against their concentration gradients. The energy derives either from the hydrolysis of ATP, in the case of proteins belonging to the ATP-binding cassette (ABC) superfamily, or from the proton motive force (PMF) across the cytoplasmic membrane. Members of the ABC family confer resistance to antiprotozoal agents and to anticancer drugs in mammals. The existence of ABC and PMF-driven transporters in fungi suggests their possible involvement in resistance to some antifungal drugs.

The major efflux systems of bacteria are energized by the PMF. The range of PMF-driven efflux systems is extraordinarily diverse and a review listed at the end of this chapter provides a detailed account of the three superfamilies of transporter proteins. Members of the major facilitator superfamily (MFS) include proteins responsible for the transport of nutritional substrates as well as for drug export. The specificity of MFS efflux proteins ranges from the low-specificity multidrug exporters described in Chapter 7 to the highly specific exporters for tetracycline antibiotics (see below). MFS transporters are large (approximately 46 kDa) proteins with either 12 or 14 transmembrane domains and hydrophilic regions that loop out both into the cytoplasm and into the periplasm of Gram-negative bacteria, and also into the (un-named) compartment external to the cytoplasmic membrane in Gram-positive bacteria. In contrast, the members of the small multidrug resistance (SMR) family, found in *Staphylococcus aureus* and other staphylococci, are believed to have only four transmembrane domains. SMR proteins frequently confer resistance to cationic antiseptics. The third superfamily of PMF-driven drug efflux proteins is referred to by the cumbersome title of the resistance/nodulation/cell division (RND) family. These proteins are thought to have 12 transmembrane domains and share an extremely broad substrate specificity. In Gram-negative bacteria the RND efflux proteins may co-operate functionally with proteins known as MFP proteins, located in the periplasmic space and outer membrane, to facilitate drug transport across both the cytoplasmic and outer membranes (see Figure 7.1). MFP proteins may also interact in a similar way with drug efflux proteins of the MFS group. The RND proteins contribute to bacterial resistance to antibiotics, antiseptics and detergents. All three families of drug efflux proteins occur in Gram-negative and Gram-positive bacteria, and members of the MFS group have also been identified in yeasts and fungi.

Examples of clinically important efflux-mediated drug resistance in bacteria, fungi and protozoa are described in the remainder of this section.

9.3.1 Tetracyclines

The clinical value of tetracyclines, which are among the cheapest and most widely used antibacterial drugs, has been severely compromised by the emergence of resistant bacteria in both Gram-positive and Gram-negative groups. There are two major mechanisms of resistance to tetracyclines, drug efflux and ribosomal protection. This account is concerned mainly with tetracycline efflux, although ribosomal protection is also described briefly.

The genes for tetracycline-specific, efflux-mediated resistance are almost exclusively plasmid and transposon located, although there is also a chromosomal tetracycline efflux system in *Escherichia coli* associated with the global regulator locus, *marA*, that enhances intrinsic resistance to many drugs (Chapter 8). High-level expression of *marA* boosts the production of the multidrug efflux pump found in many wild-type Gram-negative bacteria which includes tetracycline among its many substrates.

There is an extensive array of plasmid-borne determinants for tetracycline efflux systems among Gram-negative and Gram-positive bacteria and a review listed at the end of this chapter provides a comprehensive analysis of this complex topic. All the efflux genes code for membrane-bound proteins of approximately 46 kDa. The Gram-negative proteins have 12 hydrophobic membrane-spanning domains and the Gram-positive proteins 14. Regions of hydrophilic amino acids loop out into both the periplasmic and cytoplasmic regions. The tetracycline efflux proteins (Tet), which probably exist as multimers within the membrane, extrude a tetracycline molecule complexed with a divalent ion (probably Mg^{2+}) in exchange for a proton. Tetracycline is pumped out of the cytoplasm against its concentration gradient and the energy required for this is derived from the proton motive force. There is no direct link between tetracycline efflux and ATP hydrolysis. Gram-negative Tet proteins have functional α and β domains, corresponding to the *N*- and *C*-terminal halves of the proteins. Genetic evidence indicates that amino acids distributed across both domains participate in the efflux function. Although the Tet proteins of Gram-negative bacteria are more closely related to each other than to the larger 14-transmembrane-domain proteins of Gram-positive bacteria, the conservation of certain sequence motifs across the species suggests that the mechanisms of all tetracycline pumps driven by proton motive force are basically similar. However, the precise molecular details of how these pumps work remain to be discovered.

Regulation of tetracycline resistance

It was discovered many years ago that efflux-dependent tetracycline resistance is inducible by tetracyclines in both Gram-negative and Gram-positive bacteria. A low level of resistance is evident when the cells are first exposed to antibiotic, followed by a rapid shift to high-level resistance. The mechanism of induction is different in Gram-negative and Gram-positive bacteria. The tetracycline efflux system in Gram-negative bacteria is mediated by a structural gene for the Tet protein and by a gene coding for a repressor protein. The two genes are arranged in opposite directions and share a central regulatory region with overlapping promoters and operators. In the absence of tetracycline, α-helices in the *N*-terminal domain of the repressor protein bind to the operator regions of the repressor and structural genes, thereby blocking the transcription of both. The introduction of tetracycline complexed with Mg^{2+} leads to binding of the drug to the repressor protein. This causes a conformational change in the repressor (which has been revealed by X-ray crystallography) that eliminates its binding to the operator region and permits transcription of the repressor and structural genes. The shift from low-level to high-level resistance takes place within minutes of exposure of the bacteria to tetracycline. The process is reversible and a downshift of resistance follows the removal of tetracycline from the bacterial environment.

In contrast with Gram-negative bacteria, there is no Tet repressor protein in Gram-positive cells. The regulation of tetracycline resistance in the Gram-positive bacteria examined so far involves translational attenuation. The mRNA for the Tet protein has two ribosomal binding sites, RBS1 and RBS2. In the uninduced state the ribosome binds to RBS1 and a short leader peptide sequence is translated before the RBS2 site, which precedes the start of the structural gene proper. At this stage the RBS2 site is thought to be inaccessible within the secondary structure of the mRNA, thus preventing translation of the structural gene. The addition of tetracycline causes a conformational change in the mRNA, perhaps as a result of slowing translation of the leader sequence, which uncovers the RBS2 site. Translation of the structural region follows, leading to the expression of the tetracycline efflux pump.

Ribosomal protection against tetracyclines is mediated by the *tet(M)*, *tet(O)*, *tetB(P)* and *tetQ* genes, which are widely distributed in Gram-positive and Gram-negative bacteria, although they appear to be absent from Gram-negative enteric bacteria. In contrast with efflux-mediated resistance, the degree of resistance to tetracyclines conferred by ribosomal protection is relatively modest. The proteins encoded by the ribosomal protection genes prevent the binding of tetracyclines to their ribosomal target and are related to the elongation factors T_u (EF-T_u) and EF-G involved in protein biosynthesis.

9.3.2 Azole antifungal drugs

Azole antifungal drugs have become increasingly important as the incidence of serious fungal infections has risen sharply, especially among immunocompromised individuals. Oropharyngeal candidiasis caused mainly by *Candida albicans* is a particularly common infection in AIDS patients. Although azole antifungal agents, such as fluconazole (Figure 3.10), are generally very effective against *Candida* infections, their ever-increasing use has led to the emergence of azole-resistant strains of this pathogen. There are several modes of resistance, including a reduction in the affinity of the cytochrome P_{450} component of the 14-α-demethylase target and a substantial increase in the cellular content of this enzyme. Other resistant strains of *Candida albicans* isolated from AIDS patients fail to accumulate radiolabelled fluconazole and may overexpress either of two genes that encode multidrug efflux systems. The *CDR1* gene (signifying **C**andida **d**rug **r**esistance) encodes a member of the ABC transporter family, while the other gene, *BEN*^{r} (for **ben**omyl **r**esistance) belongs to the PMF-driven MFS superfamily. Both *CDR1* and *BEN*^{r} genes from *Candida* cloned into *Saccharomyces cerevisiae* confer resistance to fluconazole, while *CDR1* additionally confers resistance to two other drugs, ketoconazole and itraconazole. The mechanisms underlying the overexpression of *CDR1* and *BEN*^{r} in azole-resistant *Candida albicans* are unknown but could include gene amplification, mutations in the promoter regions, *trans*-acting factors, or simply a greater stability of the mRNAs. Further research is needed to establish the extent of drug efflux as the basis for azole resistance in clinical isolates of other pathogenic fungi. It is likely that decreased drug accumulation combines with other mechanisms to afford high levels of resistance to azoles.

9.3.3 Antimalarial drugs

The prevention and treatment of malaria is now in jeopardy in many parts of the world because of the resistance of malarial parasites to chloroquine, a mainstay against malaria for more than 50 years. It has been clear for some time that the resistance of *Plasmodium falciparum* to chloroquine is associated with reduced accumulation of the drug within the parasite. The discovery of two genes (*pfmdr-1* and *pfmdr-2*) for ABC-type transport proteins in chloroquine-resistant strains of *Plasmodium falciparum* suggested that the reduced accumulation of drug might be due to the activity of broad-specificity efflux pumps in removing chloroquine from the protozoal cytoplasm. However, several lines of evidence cast doubt on this appealing model:

1. Drug resistance does not co-segregate with the *pfmdr* genes in a genetic cross between chloroquine-sensitive and chloroquine-resistant strains.
2. It had been expected that the transport protein would be located in the cytoplasmic membrane in order to fulfil its putative efflux function. In fact, the protein is found mainly in the membrane of the intracytoplasmic food vacuole of *Plasmodium falciparum.* The position of the protein in the membrane suggests that it probably transports substrates, including chloroquine, into the vacuole rather than out of it.
3. Some laboratory strains of chloroquine-resistant *Plasmodium falciparum* have decreased levels of expression of *pfmdr-1* rather than the increased levels that might have been expected.
4. Increased expression of *pfmdr-1* in clinical isolates from Thailand, a region of highly resistant

malaria, is associated with resistance to two other antimalarial drugs, mefloquine and halofantrine, but increased *sensitivity* to chloroquine.

It seems unlikely, therefore, that *pfmdr*-encoded proteins have a primary role in malarial resistance to chloroquine. More recent investigations have defined another gene, *cg2*, that is closely linked to the chloroquine-resistant phenotype in *Plasmodium falciparum*. A genetic cross between a drug-sensitive strain from Honduras and a resistant strain from South-East Asia showed that the *cg2* gene derived from the resistant parasite has a complex set of polymorphisms, i.e. mutations, that are closely associated with the resistant phenotype. These specific polymorphisms were also detected in chloroquine-resistant parasites from various locations in South-East Asia and from Africa, thus strengthening the proposal that mutations in *cg2* are responsible for resistance to the drug. A different set of *cg2* mutations were identified in chloroquine-resistant *Plasmodium falciparum* isolates from South America.

The CG2 protein encoded by *cg2* is localized in the region of the cytoplasmic membrane and is also associated with haemozoin in the food vacuole of the parasite. Although the CG2 protein has no amino acid homology with known drug efflux or ion transport proteins, it may nevertheless be concerned with limiting the intravacuolar concentration and/or the action of chloroquine to subinhibitory levels. The discovery of the *cg2* gene and its protein product should provide a major impetus to research efforts to solve the puzzle of chloroquine resistance.

9.4 Other mechanisms of resistance

9.4.1 Bacterial resistance to vancomycin

Vancomycin (Figure 2.12) might well be described as the 'last chance' antibiotic because it is the only effective drug against the dangerous methicillin-resistant *Staphylococcus aureus* (MRSA) and β-lactam-resistant enterococci. Unfortunately, vancomycin-resistant enterococci are appearing in hospitals in many parts of the world. Most alarmingly, the *vanA* vancomycin resistance gene cluster in *Enterococcus faecalis* transfers to *Staphylococcus aureus* by conjugation under laboratory conditions and expresses high-level resistance. So far the threat of widespread vancomycin-resistant MRSA has not materialized but its eventual emergence seems inevitable. Details of the mechanism of vancomycin resistance are therefore of considerable importance in planning appropriate countermeasures.

The antibacterial activity of vancomycin hinges on its ability to bind to the D-alanyl-D-alanine terminus of the peptidoglycan precursor of the cell wall, thereby blocking the activity of the transglycolase and DD-peptidases essential for the synthesis of the peptidoglycan sacculus (Chapter 2). Enterococcal resistance to vancomycin rests on an unusual strategy in which the terminal D-alanine of the peptidoglycan precursor is replaced by an α-hydroxy acid, D-lactate. The affinity of vancomycin for the D-alanyl-D-lactate terminus is 1000-fold less than for the D-alanyl-D-alanine terminus of susceptible bacteria. A gene cluster harboured by transposon Tn1546 confers vancomycin resistance and encodes five proteins. VanH is a dehydrogenase that converts pyruvic acid to D-lactic acid. A ligase, VanA, catalyses the formation of an ester bond between the D-alanyl residue and D-lactate; and a third enzyme, VanX, is a DD-peptidase that hydrolyses D-alanyl-D-alanine, thereby virtually eliminating the synthesis of peptidoglycan precursors with D-alanine termini. The DD-peptidase activity of a fourth protein, VanY, removes the terminal D-alanine residue of any residual normal peptidoglycan precursors that are produced despite the attention of VanX. VanY enhances but is not essential for vancomycin resistance. The fifth protein, VanZ, confers resistance to the related antibiotic teicoplanin by an unknown mechanism.

The drastically reduced affinity of vancomycin for the terminal D-alanyl-D-lactate compared with that for D-alanyl-D-alanine stems from the elimination of a critical hydrogen bond. The amidic NH group of the D-alanyl-D-alanine linkage contributes one of five hydrogen bonds involved in the binding of vancomycin to the dipeptide. This bond is lost when the amide bond is replaced by the oxygen-containing ester link in D-alanyl-D-lactate.

The sophisticated machinery for vancomycin resistance is regulated at the transcriptional level by a two-component system. VanS is believed to be a sensor protein associated with the cytoplasmic membrane that both detects the presence of vancomycin and controls the phosphorylation of an activator protein, VanR, required for the transcription of an operon containing the *vanH, vanA* and *vanX* genes. Phosphorylation of VanR reduces its affinity for the promoter DNA of the *vanHAX* operon. A current model of the inducibility of vanocmycin resistance by vancomycin and other glycopeptide antibiotics suggests that the VanS sensor controls the phosphorylation level of VanR. The presence of glycopeptide antibiotics in the bacterial environment leads to increased phosphorylation of VanR, which in turn permits a higher rate of transcription of the *vanHAX* operon. Just how VanS detects the presence of antibiotic is unknown, but possibly an accumulation of peptidoglycan precursors caused by the antibiotic may be involved rather than a direct interaction of the antibiotic with VanS. It is thought that VanS is either a protein phosphatase or protein kinase. The model proposes that in the presence of inducing glycopeptides the activity of VanS is either inhibited or stimulated, depending on whether it turns out to be a phosphatase or kinase. In either event, the phosphorylation level of VanR would be increased.

The mechanism of vancomycin resistance described above is the one that has been most thoroughly investigated. However, it should not surprise us to find that bacteria develop other strategies of resistance to vancomycin. Already there are isolated hospital reports of vancomycin-resistant staphylococci with modes of resistance distinct from the *vanA* system.

9.4.2 Isoniazid

The efficacy of isoniazid against *Mycobacterium tuberculosis* depends partly on the intracellular activation of the compound by a bacterial catalase–peroxidase encoded by the *katG* gene (Chapter 2). While the nature of the active metabolite of isoniazid is obscure, it is clear that mutations adversely affecting *katG* confer bacterial resistance. In highly resistant clinical isolates of *Mycobacterium tuberculosis* the catalase–peroxidase activity is either absent or markedly depressed. The lack of this enzymic activity presumably results in minimal production of the toxic metabolite of isoniazid that is essential for bacterial inhibition, although this remains to be demonstrated directly.

9.5 Drug resistance and the future of chemotherapy

The development of effective drugs against microbial infections is undoubtedly one of the outstanding successes of twentieth-century science and medicine. However, as we have seen, the whole therapeutic enterprise is threatened by the relentless rise of drug-resistant bacteria, fungi, viruses and protozoa. In the case of the cellular microbes, mechanisms for eliminating toxic chemicals by compound inactivation and efflux systems existed long before the chemotherapeutic era. These, in combination with the mutability of micro-organisms, their high replication rates and, especially in bacteria, their ability to exchange and acquire new genetic material, pose enormous practical and intellectual challenges to scientists and physicians. Some authorities have raised the grim prospect of a time when the chemotherapy of infectious disease may fail completely in the face of drug-resistant organisms. However, while treatment failures already occur in specific situations, the strategy detailed below should minimize the risk of a global collapse of chemotherapy.

1. The probability of infection itself can be reduced by:

(a) insistence on high standards of hygiene in hospitals and nursing homes, in educational institutions and places of employment, entertainment and hospitality, and in the manufacture, preparation and cooking of foods;
(b) the further development and vigorous use of vaccines.

In one sense the need to treat an infection is an admission of the failure to prevent infection and many treatment regimes increase the risk of the emergence of resistant micro-organisms.

2. Undoubtedly the trivial, sometimes irresponsible use of antimicrobial drugs in human and veterinary medicine and in agriculture has assisted the spread of resistant organisms. Even the medically respectable use of antimalarial drugs as prophylactic agents to protect people in malarious zones has contributed to the emergence of drug-resistant strains of malarial parasite. Publicity and a growing awareness of drug resistance has encouraged progress towards a more restrained use of antimicrobial drugs, but their ready availability without medical supervision in many parts of the world remains a serious cause for concern.
3. The therapy of infections should be designed to ensure maximum effectiveness in terms of the choice of drug or combinations of drugs, the dose levels, dosing frequency and duration. Whenever possible the objective must be to eliminate the infecting organisms by a combination of direct chemotherapeutic attack and the activity of the patient's immune defences. The need to minimize the numbers of infecting organisms by chemotherapy is of overwhelming importance in immunocompromised patients. Exquisitely sensitive new technologies for monitoring the viral load in AIDS patients are beginning to have a major impact in managing the use of combinations of anti-HIV drugs.
4. Despite the measures listed above, there will certainly be a continuing need for new drugs to combat drug-resistant infections. Where the biochemical mechanisms of resistance are understood, ingenious drug design and skilful chemical synthesis may be expected to deliver further successes comparable with the development of novel β-lactams. The elucidation of the biochemical systems essential for microbial survival that could provide new targets for chemotherapeutic attack will continue to preoccupy many scientists in academia and the pharmaceutical and biotechnology industries. The ingenious application of molecular genetics to those micro-organisms capable of antibiotic synthesis holds the promise of creating novel structures with improved antimicrobial activities. Knowledge of the molecular genetic basis of drug resistance may also eventually be turned to good effect in devising agents to hinder the emergence and spread of resistant micro-organisms.
5. The application of agents to boost the immune defences of patients during infections has so far been relatively limited and restricted largely to immunocompromised individuals. Such treatments are generally expensive compared with antimicrobial drugs but the developing ability of biotechnology to produce the many proteins involved in immunocompetence may eventually have a broader impact on the management of infectious disease.

Further reading

Arthur, M., Reynolds, P. and Courvalin, P. (1996). Glycopeptide resistance in enterococci. *Trends Microbiol.* **4**, 401.

Bennett, P. M. and Chopra, I. (1993). Molecular basis of β-lactamase induction in bacteria. *Antimicrob. Agents Chemother.* **37**, 153.

Borst, P. and Ouellette, M. (1995). New mechanisms of drug resistance in parasitic protozoa. *Ann. Rev. Microbiol.* **49**, 427.

Bush, K., Jacoby, G. A. and Medeiros, A. A. (1995). A functional classification scheme for β-lactamases and its correlation with molecular structure. *Antimicrob. Agents Chemother.* **39**, 1211.

Chopra, I. *et al.* (1997). The search for antimicrobial agents effective against bacteria resistant to multiple antibiotics. *Antimicrob. Agents Chemother.* **41**, 497.

Cole, S. T. (1994). *Mycobacterium tuberculosis*: drug resistance mechanisms. *Trends Microbiol.* **2**, 411.

Davies, J. (1994). Inactivation of antibiotics and the dissemination of resistance genes. *Science* **264,** 375.

Ghuysen, J.-M. *et al.* (1996). Pencillin and beyond: evolution, protein fold, multimodular polypeptides and multidomain complexes. *Microb. Drug Resist.* **2,** 163.

Huovinen, P. *et al.* (1995). Trimethoprim and sulfonamide resistance. *Antimicrob. Agents Chemother.* **39,** 279.

Katz, R. A. and Skalka, A. M. (1994). The retroviral enzymes. *Ann. Rev. Biochem.* **63,** 133.

Livermore, D. M. (1995). Bacterial resistance to carbapenems. In *Antimicrobial Resistance: A Crisis in Health Care* (eds D. J. Jungkind *et al.*), Plenum Press, New York, p.35.

Molla, A. *et al.* (1996). Ordered accumulation of mutations in HIV protease confers resistance to ritonavir. *Nature Medicine* **2,** 760.

Murray, I. A. and Shaw, W. V. (1997). O-acetyl transferases for chloramphenicol and other natural products. *Antimicrob. Agents Chemother.* **41,** 1.

Park, J. T. (1996). The convergence of murein recycling research with β-lactamase research. *Microb. Drug Resist.* **2,** 105.

Paulsen, I. T., Brown, M. H. and Skurray, R. A. (1996). Proton-dependent multidrug efflux systems. *Microbiol. Rev.* **60,** 575.

Payne, D. J. (1993). Metallo-β-lactamases – a new therapeutic challenge. *J. Med. Microbiol.* **39,** 93.

Richman, D. (1994). Drug resistance in viruses. *Trends Microbiol.* **2,** 401.

Roberts, M. C. (1996). Tetracycline resistance determinants; mechanisms of action, regulation of expression, genetic mobility and distribution. *FEMS Microbiol. Rev.* **19,** 1.

Sanglard, D. *et al.* (1995). Mechanisms of resistance to azole antifungal agents in *Candida albicans* isolates from AIDS patients involve specific multidrug transporters. *Antimicrob. Agents Chemother.* **39,** 2378.

Shaw, K. J. *et al.* (1993). Molecular genetics of aminoglycoside resistance genes and familial relationships of aminoglycoside-modifying enzymes. *Microbiol. Rev.* **57,** 138.

Spratt, B. G. (1994). Resistance to antibiotics mediated by target alterations. *Science* **264,** 388.

Su, X.-Z. *et al.* (1997). Complex polymorphisms in a ~330 kDa protein are linked to chloroquine-resistant *P. falciparum* in Southeast Asia and Africa. *Cell* **91,** 593.

Thanassi, D. G., Suh, G. S. B. and Nikaido, H. (1995). Role of outer membrane in efflux-mediated tetracycline resistance of *Escherichia coli. J. Bacteriol.* **177,** 998.

Van den Bossche, H., Marichal, P. and Odds, F. C. (1994) Molecular mechanisms of drug resistance in fungi. *Trends Microbiol.* **2,** 393.

Weisblum, B. V. (1995). Erythromycin resistance by ribosome modification. *Antimicrob. Agents Chemother.* **39,** 577.

Index

D-alanine
  antagonism of antibacterial action of cycloserine 28
  liberation in cross-linking of peptidoglycan 23
  release in peptidoglycan cross-linking inhibited by penicillin 33